STATISTICAL APPROACHES TO CAUSAL ANALYSIS

THE SAGE QUANTITATIVE RESEARCH KIT

Statistical Approaches to Causal Analysis by *Matthew McBee* is the 10th volume in *The SAGE Quantitative Research Kit*. This book can be used together with the other titles in the *Kit* as a comprehensive guide to the process of doing quantitative research, but is equally valuable on its own as a practical introduction to causal inference in quantitative research.

Editors of The SAGE Quantitative Research Kit:

Malcolm Williams – *Cardiff University, UK*

Richard D. Wiggins – *UCL Social Research Institute, UK*

D. Betsy McCoach – *University of Connecticut, USA*

Founding editor:

The late W. Paul Vogt – *Illinois State University, USA*

STATISTICAL APPROACHES TO CAUSAL ANALYSIS

MATTHEW MCBEE

Los Angeles | London | New Delhi
Singapore | Washington DC | Melbourne

THE SAGE QUANTITATIVE RESEARCH KIT

Los Angeles | London | New Delhi
Singapore | Washington DC | Melbourne

SAGE Publications Ltd
1 Oliver's Yard
55 City Road
London EC1Y 1SP

SAGE Publications Inc.
2455 Teller Road
Thousand Oaks, California 91320

SAGE Publications India Pvt Ltd
B 1/I 1 Mohan Cooperative Industrial Area
Mathura Road
New Delhi 110 044

SAGE Publications Asia-Pacific Pte Ltd
3 Church Street
#10-04 Samsung Hub
Singapore 049483

Editor: Jai Seaman
Assistant editor: Charlotte Bush
Production editor: Manmeet Kaur Tura
Copyeditor: QuADS Prepress Pvt Ltd
Proofreader: Elaine Leek
Indexer: Cathryn Pritchard
Marketing manager: Susheel Gokarakonda
Cover design: Shaun Mercier
Typeset by: C&M Digitals (P) Ltd, Chennai, India

Library of Congress Control Number: 2020950514

British Library Cataloguing in Publication data

A catalogue record for this book is available from the British Library

ISBN 978-1-5264-2473-0

DEDICATION

This book is dedicated to my mother, Mary Kay McBee, who passed away before she could see this book in print.

CONTENTS

LIST OF FIGURES AND TABLES

List of figures

List of tables

ABOUT THE AUTHOR

Matthew McBee is a data scientist and educational psychologist. He has spent the last fifteen years trying to draw inferences from data and sometimes succeeding. He has worked as an applied statistician, a faculty member in a department of psychology, and most recently as a machine learning engineer. When he is not writing code, you might find him riding his bike or flying a sailplane.

ACKNOWLEDGEMENTS

This book was written in RStudio using the `bookdown` package (Xie, Allaire, et al., 2020). Most figures were made using `ggplot2` (Wickham, Chang, et al., 2020) and `DAGitty` (Textor et al., 2020). Additional R packages used in this text include the following:

- AER (Kleiber & Zeileis, 2020)
- boot (Canty & Ripley, 2020)
- broom (Robinson et al., 2020)
- cowplot (Wilke, 2020)
- data.table (Dowle et al., 2020)
- dplyr (Wickham, François, et al., 2020)
- gbm (Greenwell et al., 2019)
- ggdendro (de Vries & Ripley, 2020)
- gridExtra (Auguie & Antonov, 2017)
- ivmodel (Kang et al., 2020)
- knitr (Xie, Vogt, et al., 2020)
- MASS (Ripley et al., 2020)
- Matching (Sekhon, 2019)
- MatchIt (Ho et al., 2018)
- multcomp (Hothorn et al., 2019)
- png (Urbanek, 2013)
- PSAgraphics (Helmreich & Pruzek, 2012)
- psych (Revelle, 2020)
- rdd (Dimmery, 2016)
- reshape2 (Wickham, 2020)
- rockchalk (Johnson & Grothendieck, 2019)
- rpart (Therneau et al., 2019)
- survey (Lumley, 2020)
- tidyr (Wickham & RStudio, 2020)
- truncnorm (Mersmann et al., 2018)
- twang (Ridgeway et al., 2020)

PREFACE

I first encountered many of the concepts of this book circa 2008. I was working as an applied statistician at UNC-Chapel Hill at the time. The statisticians' group I belonged to would read and discuss papers for professional development, and that year we read several of the causal inference 'classics' from Campbell, Pearl, Rubin and others which are heavily cited in this book. It was a fortuitous and formational event in my career, though it took time for some of the ideas to take root. Later that year, my colleague Samuel Field and I spent hours over lunch at the white board working through the concepts of propensity score analysis. We ran simulations, sketched graphs, read papers, and derived equations together until we were thoroughly convinced. I still believe that this is the best way to learn new statistical ideas.

I later took a faculty job in quantitative psychology at East Tennessee State University. I was responsible for teaching most of the statistical methods courses for our experimental and clinical psychology PhD students. Early into my role as a faculty member, the severity of the replication crisis in psychology became increasingly clear as one large-scale replication study after another failed to support original claims. Certain phenomena that seemed to be established beyond a reasonable doubt and supported and extended by dozens or hundreds of studies fell apart under the harsh light of preregistered experiments. It was clear that the usual practices of data analysis, study design, and dissemination of results were not up to the task of facilitating true scientific progress.

The epicenter of the replication crisis in psychology was in the subdiscipline of social psychology, where the most common research design is a randomised lab experiment. Meanwhile, in my field of educational psychology, experiments are comparatively rare. The most common designs in my field are quasi-experimental or observational studies relying on linear regression (or a related technique) for inference. If social psychology had gone so wrong with experiments, I wondered how well my own field could possibly have fared under similar incentives and statistical training but weaker designs. Unfortunately, the answer to this question is still unknown at the time of this writing.

Experiencing the replication crisis in psychology from such a close vantagepoint made me keenly aware of the promise and especially the peril of statistics. The ease with which people can fools themselves with statistics is something that has long been appreciated in an abstract sense, but the replication crisis gave this idea a deep and abiding reality in my mind. And the risks of fooling oneself are much higher when the design of a study itself provides little or no protection against certain inferential errors. And thus, my interest in the methods of causal inference springs from the same source as my dismay regarding the replication crisis. Just as a 'p-hacked' statistical analysis is likely to lead to an incorrect conclusion, an uncareful regression of observational data is also likely to mislead. The methods of causal inference discussed in this book are not magic truth algorithms; their value is in facilitating careful and informed decision-making and providing additional diagnostics to help identify problems that would otherwise remain invisible.

The most difficult aspect of analysing observational data with regression models is determining what covariates need to be included in the model in order to produce estimates that have at least a vaguely causal interpretation. This is among the most important of decisions an analyst must make, and yet traditional books and courses on linear regression provide very little guidance on how to make it. Directed acyclic graphs, which provide an organising principle for this book, provide a principled means of making this critical choice. Directed acyclic graphs therefore became central to my thinking about research as well as a focus of my teaching in both undergraduate and graduate courses. They fill a critical need in the researcher's toolbox. I hope they are one day a cornerstone of statistical education.

When series editor Betsy McCoach approached me about the possibility of contributing to the *SAGE Quantitative Research Kit* series, I had just finished teaching a graduate seminar on casual inference. Like so many instructors before me, I reflected on the work I had done to prepare for the class and thought to myself that it would not be so hard to flesh out those notes, materials, and examples into a book. This was wrong, of course, but many books would probably not exist if not for this common delusion. This book is no exception.

During the development of this book, I entered a new phase of my career, where I became a data scientist working in private industry. Though the scope of problems I encounter are different, the principles and perils are the same. It is one thing to train a machine learning algorithm to make predictions, which can be happily accomplished with no notion of causality at all. But it becomes entirely different to construct a model that will accurately represent what might happen if inputs are perturbed in a systematic fashion. For example, if business practices are altered such that certain regressor variables change in a certain isolated way, how would we expect the outcome variable to change in response? Posing that question moves the context from the world of seeing to the world of doing; from correlation to causal inference. While the examples given throughout this book are set in the context of a researcher performing a scientific investigation, the principles described have broader application.

I hope you enjoy reading this book and find it useful in supporting your own inferential efforts, whatever they might be. I am grateful to the series editors and to SAGE for the opportunity to write it.

Matthew McBee

1

INTRODUCTION

Chapter Overview

In my experience, students of the social sciences receive much more instruction regarding what causation is not than what it is. When asked 'what is causation?' the first answer given by the graduate students in my linear modelling course was, 'Correlation does not imply causation!' This is correct as a general principle, of course. But causation nearly always implies correlation. There are conditions under which a correlation, slope coefficient or group mean difference measures the magnitude or strength of the causal relationship between variables. But there are many others in which these summary statistics imply nothing of the sort. In some conditions, these statistical values become fine instruments allowing us to peer into the heart of phenomena; in others, they are nearly as useless as the astrological signs or creases on a palm. One of the great challenges of empirical science is the proper understanding of when observed statistics can be imbued with meaning and when they should be ignored. Correlation does not imply causation, except for when it does.

In psychology and education, the fields I was trained in, the dominant framework for understanding causal inference and its connection to research design comes primarily from Campbell and Stanley's (1966) classic, *Experimental and Non-Experimental Designs for Research*, and the later work that it inspired, such as Shadish et al. (2002). Campbell and Stanley's short and readable monograph is freely available online[1] and is well worth careful study, as it introduced a number of concepts that are now central to social research.

Internal validity

Campbell and Stanley (1966, p. 5) described **internal validity** as 'the basic minimum without which any **experiment** is uninterpretable: Did in fact the experimental treatments make a difference in this specific experimental instance?' They went on to describe it as the *sine qua non* of research. To conclude that a study exhibits internal validity is equivalent to determining that it supports causal inference. As this book is primarily interested in the question when and how researchers can draw causal conclusions from data, one could argue that this book is primarily concerned with internal validity.

External validity

External validity describes the degree to which a **causal effect** is generalisable to other populations, contexts or settings. In social research, controlled laboratory

---

[1]www.sfu.ca/~palys/Campbell&Stanley-1959-Exptl&QuasiExptlDesignsForResearch.pdf

experiments are often portrayed as high in internal validity but low in external validity, because it is unclear to what degree behaviours exhibited in the artificial environment of a lab mirror those exhibited in real-life contexts. On the other hand, field experiments may have high external validity but suffer from low internal validity due to the lack of control researchers have over extraneous factors in such settings (Shaughnessy et al., 2015). Thus, internal validity and external validity are in tension. Studies relying on animal models may have the highest baseline internal validity imaginable due to the researcher's ability to control the environment, diet, exercise, social exposure and even the genetics of the animals. However, the external validity of such studies is questionable due to the obvious differences between animal and human physiology and behaviour.

External validity concerns appear in this book in several places. Propensity score analysis (which is introduced in Chapter 5) allows researchers to estimate causal effects that apply to different subgroups of the population. But the most important concerns about external validity arise in the context of instrumental variables analysis (Chapter 6) and regression discontinuity designs (Chapter 7). These techniques estimate causal effects for very specific subgroups. The exact identity of those groups may be unclear, or they may not be exactly the population to which the researcher hopes to generalise. These subpopulation-specific causal effects are called *local average treatment effects*. The general inability of instrumental variables and regression discontinuity design techniques to estimate *global* causal effects is a limitation of those methods.

Threats to validity

A **threat to validity** is an alternative, non-causal explanation for apparent differences between group means or other suggestive quantitative indications of potential causal link between variables. Campbell and Stanley (1966) described eight threats to internal validity, which are displayed in Table 1.1. They also described four threats to external validity.

Table 1.1 Threats to internal validity

Threat	Description
History	Occurrence of outside events during the course of a study
Maturation	Natural developmental change during the course of a study
Testing	Practice effects: later performance affected by prior assessment

(Continued)

Table 1.1   (Continued)

Threat	Description
Instrumentation	Changes in measurement or calibration of instruments
Regression to the mean	Tendency for extreme scores to shrink upon repeated measurement
Selection	Systematic differences between subjects assigned to groups
Experimental mortality	Differential study dropout from the groups
Selection–maturation interaction	Systematic differences between groups cause differing rates of development

From Campbell and Stanley's perspective, the quality or rigour of a study is proportional to the number of these threats that can be logically eliminated or rendered improbable by virtue of its design. Causation is provisionally accepted after counter explanations have been considered and rejected. Shadish (2010) described Campbell and Stanley's (1966) perspective as rooted in the Popperian falsificationist philosophy of science. In Popper's philosophy, an experimental result can only falsify or fail to falsify a theoretical proposition. It is not possible for an empirical finding to verify that a theory is correct, or even to directly support a theory per se. It can only show that the theory is incorrect or incomplete in some way when it fails to account for an empirical observation (Dienes, 2008). The best experimental outcome that a theorist can hope for is survival. As Campbell and Stanley (1966) wrote,

> In a very fundamental sense, experimental results never 'confirm' or 'prove' a theory – rather, the successful theory is tested and escapes being disconfirmed. The word 'prove', by being frequently employed to designate deductive validity, has acquired in our generation a connotation inappropriate both to its older uses and to its application to inductive procedures such as experimentation. The results of an experiment 'probe' but do not 'prove' a theory. An adequate hypothesis is one that has repeatedly survived such probing – but it may always be displaced by a new probe. (p. 35)

In a similar fashion, Campbell and Stanley do not affirmatively define the conditions under which a true causal relationship can be concluded. Instead, they described conditions, circumstances and events that would create the illusion of causation when it does not exist. Consider their description of the 'One-Group Pretest–Post-test Design', in which subjects are measured on some variable, exposed to a 'treatment' and then measured again. Suppose that the outcome variable is the severity of depression symptoms and the treatment is cognitive behavioural therapy. Further, suppose that the mean depression symptoms at post-test are substantially (and statistically significantly) reduced from their mean value at pretest. Shall we conclude that the therapy caused a reduction in symptoms?

Campbell and Stanley (1966, Table 1, p. 8) indicate that this design is weak regarding the following threats to validity:

- *History*. It is possible that some outside event took place between the pretest and post-test measurements besides exposure to treatment which affected the subject's depression symptoms. For example, perhaps the pretest took place in the winter and the post-test in the summer, and the increase in light, pleasant weather and outdoor activity affected the subjects.
- *Maturation*. Depression symptoms may spontaneously resolve without treatment for some subjects. This could be mistaken for an effect of treatment. More generally, natural change over time as a result of development or other processes can alter post-test scores independent of the treatment.
- *Testing*. The act of completing the first depression assessment may itself alter responses to the second assessment even if the subjects' underlying level of depression has not changed. This threat may seem to be a remote possibility in this example, but if the assessment was performance based, the subjects may do better on the post-test assessment due to their experience of taking the pretest.
- *Instrumentation*. Calibration of measurement instruments may drift over time, or the instruments may fail, creating an illusory apparent change in scores from pretest to post-test. This once happened to me when a device for recording heart rate began to fail in the midst of data collection. This may seem unlikely if the instrument is a self-reported symptoms rating scale that cannot fail or lose calibration, but it is possible for even simple survey instruments to exhibit problems with longitudinal measurement invariance, in which the meaning or value anchoring of scores drift over time (Meade et al., 2005).
- *Regression*. When subjects are selected on the basis of extreme scores at baseline, it is likely that follow-up measurements will be less extreme. This is an unavoidable consequence of measurement error (Crocker & Algina, 1986). Thus, the post-test depression scores may improve even if the treatment is ineffective.
- *Mortality*. 'Mortality' refers to dropping out of the study. If there is an association between depression severity and the probability of dropping out, or an association between response to treatment and the risk of dropping out, then a comparison of pretest and post-test means will not necessarily represent the causal effect of the treatment. For example, if the most severely depressed subjects quit the study, the post-test mean symptoms score will improve simply because the subset of subjects remaining is less depressed, even if the treatment had no effect.

Campbell and Stanley (1966) describe several alternative research designs that are much more robust to these threats to validity than the 'One-Group Pretest–Post-Test Design' considered above. One of these is called the 'Pretest–Post-Test Control Group Design'. In this design, subjects are randomly assigned to a treatment group and a control group. Both groups receive a pretest assessment. The treatment group is exposed to the active treatment condition and the control group to a control or placebo condition. Then both groups are assessed again at post-test. The treatment effect in this design is measured by the difference in pretest to post-test change for the treatment group versus the control group. Any change resulting from history, maturation, instrumentation or regression should affect both groups equally and will not systematically impact the estimated treatment effect. Mortality is still a potential problem (e.g. if the treatment is unpleasant and induces people to quit the study), but if there is little to no dropout, it becomes an implausible alternative explanation.

Randomisation

The randomisation of treatment assignment in this design is an essential feature. Randomised experiments were an incredible innovation that are widely attributed to Fisher (1935, 1936). Instead of attempting to hold every factor constant across conditions except for exposure to the treatment, randomised allocation of subjects to treatment conditions 'controls for nothing' (Guo & Fraser, 2015), ensuring that the treatment and control subjects are equivalent in expectation (i.e. in the population) on all background factors. (Interestingly, this idea of holding background factors constant as an alternative to randomisation has made a comeback. It is now one of the front-line modern strategies of causal inference and is the topic of Chapters 2–5 of this book.) Randomisation removes the selection threat to validity: the baseline unequivalence of groups exposed to the levels of the variable being studied. Campbell and Stanley (1966) presented a number of clever non-randomised (or quasi-experimental) designs tailored to rule out various threats to validity. They summarised the strengths and weaknesses of these non-randomised designs (see Table 2, p. 40), showing how some protect against history effects, others protect against maturation and so on.

Surprisingly, Campbell and Stanley (1966) describe several of these quasi-experimental designs as offering strong protection against the selection threat, which is initially somewhat surprising until one appreciates exactly how they considered allocation to treatment to occur under their conception of a quasi-experiment. For example, their 'Nonequivalent Control Group Design' (p. 47) involves taking two (or more) pre-existing groups of subjects (e.g. classrooms), assessing the students within them, *randomly* exposing some of these groups to treatment, the others to control, then assessing them again. They view this as a non-randomised design because the researcher did not assign subjects to groups even though he or she assigned the groups to levels of treatment. This would now be considered a *cluster-randomised design* (Raudenbush, 1997) with strong causal inference features due to randomisation. Indeed, a modern perspective would consider this design somewhat troublesome statistically (as classical models do not account for the resulting dependence of subjects in groups) but quite straightforward regarding internal validity or causal inference.

Campbell and Stanley's (1966) work promoting quasi-experimental methods can be understood as a reaction to the dominance of the dogmatic experimentalist approach to psychology and education research that defined those fields from the 1930s through the 1960s (Shadish, 2010). They defined quasi-experiments as those designs in which the researcher has at least *partial* control over the assignment or timing of presentation of experimental conditions, writing,

> One dimension of 'quasi-ness' [emphasis in original]. . . is the extent to which the X could be manipulated by the experimenter, i.e., could be intruded into normal course of events. Certainly, the more this is so, the closer it is to true experimentation. (Campbell & Stanley, 1966, p. 64)

It is a mark of Campbell and Stanley's success that modern readers view many of these initially controversial designs as having the essential features of experiments.

Non-experimental research

Campbell and Stanley (1966) considered a third form of research: the correlational and ex post facto designs, which are also known as observational studies. These have the defining characteristic that the researcher has no control at all over the presentation of the experimental condition. These they viewed with great scepticism. Of the general correlational design, they wrote, 'we are left with the general rule that the differences between two natural objects are uninterpretable' (p. 64), which they attributed to the possibility that other factors besides treatment exposure which vary between groups could very well be responsible for any differences observed between them. These competing explanations are called confounders, and the disturbance they exert on summary statistics is known as confounding (Brookhart et al., 2010).

They described correlational studies as producing an asymmetrical form of evidence for which only values of zero are interpretable. While a non-zero correlation[2] is uninterpretable due to confounding, a zero correlation between variables that are hypothesised to be causally connected rules out such a relationship. Correlation does not imply causation, but a lack of correlation does indicate the absence of causation

Fundamental progress in the theory and practice of causal inference has occurred in the years since 1966, and also in the context in which a great deal of social research

---

[2]In this context, I am using the word 'correlation' in a generic sense to refer to any type of apparent relationship between variables – not the strictly linear relationships described by the Pearson correlation coefficient.

takes place. Whereas the Platonic form of social research (at least in psychology and education) in the mid-1960s might have been a classical randomised experiment, the explosion of accessible data enabled by advances in computers and networking technology has firmly established correlational research as the dominant type in many social science disciplines. Fields such as astronomy and cosmology have made tremendous progress despite a complete lack of intervention or manipulation of core constructs. So have fields such as economics, epidemiology and personality psychology, in which experiments are comparatively rare. Modern theories of causal inference provide a way forward into the frontier of the correlational or observational designs that Campbell and Stanley regarded as essentially hopeless. They describe the conditions under which even non-zero statistical quantities can be interpreted as meaningful and provide guidance on how to achieve those conditions. Modern theories do not supplant Campbell and Stanley. History effects, practice effects, regression effects and all the others must be guarded against. But preventing these is a relatively trivial problem compared with the challenge of making valid inferences when confounding is likely.

When possible, randomised manipulation of variables is preferred under most circumstances. Randomisation removes the selection or confounding threat by ensuring that the treatment and control subjects are equivalent in expectation (i.e. in the population) on all background factors. But even randomised experiments can fail. Subjects may not perfectly comply with their assigned treatment – some will forget to take the pill, refuse to adhere to the diet or will stop attending therapy sessions. Some subjects in the control group may obtain the treatment. And some subjects might drop out of the study before it is complete. In such conditions, treatment may be randomly assigned, but treatment compliance is not. The confounding threat is reasserted. These 'broken' randomised experiments have become quasi-experiments.

This book considers and introduces some modern ways of thinking about and assessing causal relationships between variables in observational studies, which rely on processes out of the researcher's control to produce variation in the variable of interest. For example, researchers might study the relationship between family financial resources and children's academic achievement. Wealth is highly variable in the population, enabling researchers to calculate a measure of association (e.g. a correlation coefficient or regression slope) between these variables. Observational designs are profoundly exposed to the confounding or selection threat, as students from different levels of family wealth differ on many other dimensions as well. Alternatively, researchers might look for situations in which wealth changes suddenly; perhaps due to an unexpected windfall such as an inheritance or winning the lottery, or perhaps due to a job loss or policy change affecting incomes. This sudden change, called an 'exogenous shock' by economists, can help to mitigate the selection threat. Chapter 6 of this book discusses how these shocks can be exploited to estimate causal effects.

A pragmatic definition of causation

This book adopts Pearl et al.'s (2016) pragmatic definition of the term: X causes Y if Y depends upon X for its value (p. 5). This definition encompasses the three aspects of John Stuart Mill's conditions for inferring causation: (1) that the cause precedes the effect, (2) that the cause is related to the effect and (3) that we can find no alternative explanation for the effect other than the cause (Shadish et al., 2001). Pearl's definition indicates that if X causes Y, then manipulating or altering the value of X would create a corresponding change in Y, but not the opposite: altering Y through other means would not create change in X. Thus, causation is directional and temporally sequenced, thus fulfilling Mill's first two conditions.

Mill's third condition is not to imply that X is the only cause of Y, or that all changes in Y must be accompanied by changes in X, but is simply a matter of attribution. If we observe a change in X followed by a change in Y, the change in Y must occur because of X, not because of simultaneous change on A, B or C. Mill's third condition can be satisfied by *isolating* change in X such that X alone, and no other factor, is being altered at the same time. Most social, psychological, medical or economic events are multifactorial in nature with complex and often interactive etiology. This complexity does not contradict the notion of causation. Just because event Y can sometimes occur in the absence of X does not imply that X is not a cause of Y. The fact that infections can resolve without medication does not imply that antibiotics have no effect. Mill's third condition simply means that, if we wish to determine whether antibiotics cured the infection, we must not administer them in combination with other potential treatments.

Making causal inferences outside of randomised experiments has taken on a near-taboo status in certain fields of the social sciences, in spite of the fact that only a small proportion of the work in those fields is experimental. This has given rise to the talisman-like disclaimers that are found in so many non-experimental journal articles, in which authors disclaim any notion that their findings provide evidence of causation. Taking such statements seriously raises serious questions about the value or utility of the work. Luckily, one cannot seriously entertain the notion that the authors themselves hold their own work to be meaningless. The abstract, introduction, discussion and implications sections of such manuscripts are almost always leveraged on the causal interpretation that the authors forbear. Science is not the law, reality isn't a jury and disclaimers are of little use when they are flatly contradicted by implication. Warning against causal interpretation has become the small print of social science.

But scientists are going to make causal inferences. We must. We might as well do so as safely and wisely as possible.

Prediction versus explanation

One of the most fundamental goals of science is the accurate prediction of future events. Predictions may be made on the basis of causal or non-causal associations. For example, a researcher may be able to predict the rate of heat stroke from ice cream sales and other non-causal variables, perhaps with high accuracy. But understanding the actual *causes* of heat stroke offers three advantages: (1) precision, (2) parsimony and (3) intervention leverage.

Precision. Non-causal associations exist because they are noisy proxies of the underlying causal relations. Heat stroke is not caused by eating ice cream; both are driven by prolonged exposure to high temperatures. If predictions can be derived from noisy proxies, better and more precise predictions can often be made from the causational drivers of events. No proxy variable can have a stronger relationship with the phenomenon of interest than its causes.

Parsimony. Parsimony is a strong heuristic preference in science. Given a set of competing accounts of a phenomenon, all of which produce equally accurate predictions, the simplest is favoured. This principle is justified not only on the basis of falsifiability but also in terms of raw efficiency. After all, if the goal is to predict heat stroke, one could imagine accounting for a huge set of related but non-causal observations: the number of boats in the lake, the proportion of convertibles with their tops down, the proportion of people wearing sunglasses, sunscreen sales, the number of cars coming in for AC (air conditioning) repairs and anything else you can imagine. But given that measurement is not free, but always involves effort and expense, it would be far more efficient to base predictions off the much smaller set of variables that actually *cause* people to experience heat stroke: outside temperature, humidity, cloud cover, proximity to air conditioning, underlying health and so on, even if the predictions derived from each were equally precise.

Intervention leverage. Most domains of the social sciences have a strong applied focus. The goal is not only to predict if and when behaviours or phenomena occur but to be able to alter them. We would like to know how to best bring an end to economic recessions, effectively treat anxiety disorders, prevent mass shootings, resolve intergroup conflict, reduce stigma and increase academic achievement. Non-causal proxy variables may enable prediction, but they are useless for intervention. After all, policies preventing people from buying ice cream or from wearing sunglasses – both factors that may *predict* heat stroke, but do not *explain* it – would not be very effective.

Causal inference requires external information

The coefficients produced by statistical analysis are always fundamentally transformations of the correlation coefficient or the semi-partial correlation coefficient. See Table 1.2 for some examples of how common effect sizes from various statistical models can be converted into correlations. They are correlations in disguise. And bearing in mind the statistician's catechism that correlation does not imply causation, the implication is that *no* summary statistic in itself implies causation, regardless of whether that correlation masquerades as a Cohen's *d* effect size, an odds ratio, a regression slope coefficient or a structural path coefficient in a latent variable model.

Table 1.2 Some common effect sizes and their transformation to correlation coefficients

Model	Statistic	Transformation to r
t-Test[a]	Cohen's d	$\sqrt{\dfrac{d^2}{4+d^2}}$
ANOVA[b]	Eta squared	$\sqrt{\eta^2}$
Simple regression[c]	Slope coefficient	$b\left(\dfrac{sd(X)}{sd(Y)}\right)$
Multiple regression[c]	Slope coefficient[d]	$b\left(\dfrac{\sqrt{var(X)\left(1-R_x^2\right)}}{sd(Y)}\right)$
	Model r^2	$\sqrt{r^2}$
Logistic regression[e]	Odds ratio	$\log(OR)\left(\log(OR)^2+\dfrac{2.89\,n^2}{n_1\,n_2}\right)^{-1/2}$

Note. ANOVA = analysis of variance; OR = odds ratio.

[a]Equation given by Borenstein et al. (2011), assumes equal group sizes.

[b]Equation given by Lakens (2013).

[c]Equations given by Fox (1997).

[d]Equation for the semi-partial correlation of Y with X controlling for all other predictors in the model. R_x^2 is the R^2 statistic from a model regressing X on all the other predictors – the tolerance.

[e]Equation is Ulrich and Wirtz's (2004) approximation to the point-biserial correlation. n_1 and n_2 are the number of cases in each level of the binary response variable.

Inferring causation always requires information external to data. Pearl et al. (2016) emphasised this point in the introduction to their excellent primer on causal inference. The interpretation of data requires some theory about the data-generating process. The level of detail and nuance required of this external information is directly proportional to features of the design. For example, knowing that the treatment variable is randomly assigned means that one needs no additional information in order to interpret the data, assuming that all subjects fully comply with their treatment assignment and that there is no missing data. What if there is missing data? Now additional information must be incorporated – information about the process responsible for the missing data. Was the missing data caused by a failure of data collection equipment, or was it caused by subjects dropping out of the study? If the latter, is it possible that the probability of dropping out is related to whether the subject was experiencing benefit or harm from the treatment? If so, the causal effect of the treatment cannot be extracted from the data unless the researcher commits to a model of the missing data mechanism – a model that cannot be evaluated except by reference to the very information that is missing. In this case, a raw correlation coefficient – or any of the other 'correlation in disguise' summary statistics from Table 1.2 – are potentially misleading and incorrect descriptions of the causal effect of the treatment.

Observational designs require the most detailed and encompassing auxiliary information of all. *Assumptions* are the proper name for this auxiliary information, which does not come from the data itself, but is nonetheless required in order to properly analyse the data and to interpret the result. Chapter 3 of this book describes a formal system, called directed acyclic graphs, for clarifying, communicating and making modelling decisions based on assumptions about the data-generating process. At times it may seem impossible to know all that must be known to credibly make causal inferences from observational studies. I will not argue against this position; the challenge of inference in such studies is extraordinary. Still, many of the most interesting and most critical questions that can be approached by the social sciences are ones in which manipulation (experimental or otherwise) of the key variable would be impossible or unethical. Limiting empirical social science to experimental study alone would serve to vastly limit the scope of inquiry in which we could engage and the degree to which the behavioural sciences could progress. Entire topics of study, including intelligence, personality, behavioural genetics, sexual orientation, gender identity and many others, would simply fall outside the boundaries of empirical investigation. These issues demand inquiry, and so we must try to investigate them using the methods that are available to us.

When causal assumptions are met, mere descriptions of data become something much more. They illuminate the process giving rise to the data. They become accurate measures of the strength of causal connections between variables. They provide a link from the world of 'seeing' into the world of 'doing' (Dawid, 2010), allowing researchers to infer the consequences of manipulating the systems they study. However, causal inference is challenging, particularly outside the context of carefully controlled experiments. Scientific progress is not assured, but it is possible; and that provides a sufficient justification for the effort. It is best to approach the goal of causal inference with a healthy respect for the challenge involved and a thorough understanding of the assumptions that such interpretations are leveraged upon. It also pays to be humble and sceptical; one may well arrive at a wrong answer in spite of one's best effort.

When these causal assumptions are violated, there are no error messages and generally no warning signs. The software produces output. The process fails silently. There are some methods for assessing the plausibility of assumptions, but all are incomplete. They can eliminate some but not all the potential mistakes that researchers can make. It is therefore best to rely on methods that rely on the smallest set of assumptions possible, to assess the validity of those that remain as rigorously as possible and, most of all, to clearly and honestly disclose the details of the research process. In so doing, the self-correcting nature of science will lead to incremental progress.

Estimation versus hypothesis testing

Analysts may sometimes be more interested in the question of whether one variable causes another, and other times more interested in the magnitude of the relationship (and the precision with which it was estimated). The question of *whether* is addressed by hypothesis testing; the question of *magnitude* by point estimation with uncertainty bounds. This book focuses on the estimation perspective. For example, results will often be presented as an estimate and 95% confidence interval [CI]. The reader will be reminded of what the true value for the parameter should be in order to clearly illustrate the performance of these methods under various conditions. This presentation provides the most straightforward connection to the issues of bias, variance and consistency that are fundamental to this discussion. Estimation (as opposed to hypothesis testing) tends to be emphasised in the literature on causal inference. The question of whether a causal effect is present or absent is subsumed in the estimate of its magnitude; I regard the magnitude question as simply more informative (Cumming, 2013). But estimation and hypothesis testing are intimately linked; any factor that disturbs the accuracy or precision of estimation will also affect the error rates of statistical tests and the coverage rate of confidence internals (DeGroot & Schervish, 2002).

Prerequisites

This book assumes that readers have a solid grounding in basic inferential statistics and especially in linear regression analysis. Those desiring a review of this material are directed to Volumes 3 (*Statistical Inference and Probability*) and 8 (*Linear Regression: An Introduction to Statistical Models*) of this series.

Notation

Table 1.3 summarises the mathematical notation used throughout this book. I only emphasise the vector or matrix identity of variables when I deem that doing so increases clarity. For example, I typically refer to random variable X instead of $\mathbf{X}$ because it should be clear from the context that X contains more than one value.

Table 1.3 Notation

Bold font	A matrix or vector element
Capital italicised letters (X, Y, Z)	Random variables
X_i	The value of X for the ith subject
$\mathbf{X}$	The set of control variables or covariates
$X \perp Y$	X and Y are independent
Greek characters (α, β, γ)	Population parameters
n	The sample size
D	Treatment status represented as a {0, 1} indicator variable
Y	The outcome or response variable
Y_0, Y_1	The potential outcomes under control and treatment conditions
U	An unobserved or unobservable variable
$\hat{Y}$	The predicted values of Y, 'y-hat'
$E(X)$	The marginal (e.g. overall) expected value of X; its mean
$E(X \mid Y = y)$	The conditional expectation of X given Y has value y
$p(X = x)$	The marginal probability that random variable X takes value x
$p(X = x \mid Y = y)$	The conditional probability X takes value x given that Y has value y
$sd(Y)$	The standard deviation of Y
$var(X)$	The variance of X
cov_{XY}	The covariance between X and Y
r_{XY}	The estimated Pearson correlation between X and Y
ρ_{XY}	The population correlation coefficient between X and Y
$X \sim N(a, b)$	X follows a normal distribution with mean a and standard deviation b

The R statistical programming environment

This book uses the free statistical programming language R for all the examples. However, it is not an 'R book' in that it does not focus on how to write code to perform the various analyses that are described. At times, I will give some brief description of the steps I took in the software to perform various computations, but there is no R code in this book. There are, however, code and data available on the book's companion website for download so readers can re-create the analyses in this book. This book frequently displays R output to serve as a bridge between the abstract ideas being discussed and the practical matter of reading and interpreting statistical package output. The analyses presented in each chapter are applied to simulated data sets. The advantage of simulated data is that the true values of the parameters are known with certainty, making it easy to explore the boundaries of these methods – when they work, and when they fail. It is quite difficult to empirically demonstrate concepts such as bias and consistency with real data because one never knows the ground truth.

Verbatim R output is displayed in a fixed-width font with each line preceded by ##, like this.

```
##
## Call:
## lm(formula = mpg ~ hp, data = mtcars)
##
## Residuals:
##      Min      1Q  Median      3Q     Max
## -5.7121 -2.1122 -0.8854  1.5819  8.2360
##
## Coefficients:
##              Estimate Std. Error t value Pr(>|t|)
## (Intercept) 30.09886    1.63392  18.421  < 2e-16 ***
## hp          -0.06823    0.01012  -6.742 1.79e-07 ***
## ---
## Signif. codes:  0 '***' 0.001 '**' 0.01 '*' 0.05 '.' 0.1 ' ' 1
##
## Residual standard error: 3.863 on 30 degrees of freedom
## Multiple R-squared:  0.6024, Adjusted R-squared:  0.5892
## F-statistic: 45.46 on 1 and 30 DF,  p-value: 1.788e-07
```

R is a free, open-source and highly extensible software application, and certainly one of the best tools available for applied statistics. I recommend that R users also download RStudio, which is also free and open source. RStudio is an *integrated development environment* for R that makes it easier to use and adds some nice features, such as better workspace organisation, as well as the ability to easily incorporate code, statistical output and plots into documents, presentations and interactive web apps. These programs are available for Windows, Mac and Linux machines, and they are lightweight, so they will work on old computers quite well.

Installing and using R and RStudio

R should be downloaded and installed first. You can download R from www.r-project.
org. RStudio can be downloaded from www.rstudio.com.

Figure 1.1 is a screenshot of RStudio. Code can be entered into the console (lower
left), where it is run line by line, or into the script editor (*File → New File → New Script*)
in the upper left. Script editor code is run by highlighting it and clicking *Run*. Textual
output appears in the console; plots appear in the lower right window, which is also
where help files are displayed.

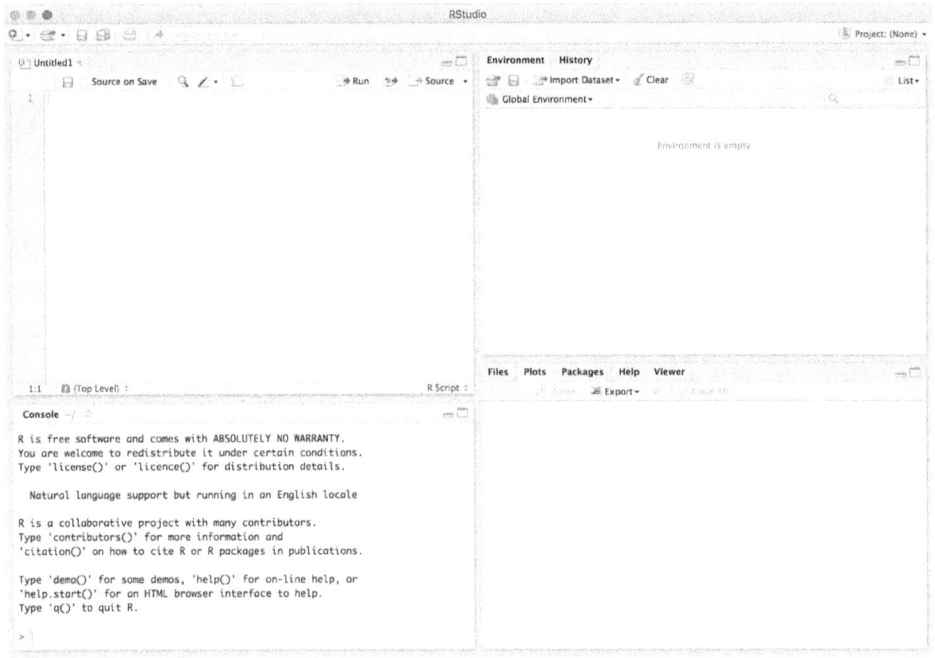

Figure 1.1   RStudio

The help file for any R function can be viewed by entering a question mark followed by the function name at the console. For example, to view help for the rnorm
function, type

```
?rnorm
```

R packages

Many of the examples in this book make use of functions that are not included in the
base R but are distributed in free, user-developed downloadable add-on packages. A
two-step process is required to obtain and activate these packages.

Step 1: Download the package

The survey package is used in Chapter 2 to do weighted regression analysis. The first step in obtaining this package is to download it to your computer. Many R packages are hosted on *CRAN* (the 'Comprehensive R Archive Network') and can be installed directly from the R console using the install.packages() function. This step only needs to be done once. The following line of code will download the package. Note that the package name must be enclosed in quotes, and that R code is case-sensitive.

```
install.packages("survey")
```

Step 2: Attach the package

Once the package has been downloaded, it needs to be loaded. This makes its functions available to be used and is done with the library() or require() functions. This function needs to be rerun each time you restart R. Note that the package name does not need to be quoted in the library() function.

```
library(survey)
```

Now the functions from the survey package are available for use.

In summary, R is simply a means to an end. Any statistical software with functions to implement the methods discussed in this book will produce similar results to the R output that is shown, with allowance for the idiosyncrasies of output formatting across software. Stata in particular has excellent routines for implementing the techniques that will be covered here. Many of these methods are implemented in other statistical software applications and programming languages such as SAS, SPSS or Python. It is my hope that readers preferring other software will not be alienated by the R output and will have little difficulty in understanding and translating it into their preferred style.

Structure of this book

The most important concept in causal inference is the idea of conditioning – incorporating information about auxiliary variables into an analysis. Chapter 2 reviews various model-based and non-parametric conditioning methods. Chapter 3 describes directed acyclic graphs, which provides an organising schema for considering the issues presented in the remainder of the book. Chapter 4 provides an overview of Rubin's causal model, also known as the **potential outcomes** framework, which furnishes a clear definition of a causal effect and provides a mathematical framework for understanding causation and the conditions under which it can be inferred. It introduces the concept of propensity scores, which follow directly from Rubin's model, and describes how they can be estimated from data. Chapter 5 discusses three methods for using propensity scores to perform causal inference: matching, subclassification and weighting. Chapter 6 describes instrumental variables analysis, a technique enabling causal inference even when much less is known about all of the variables that must be controlled. Chapter 7 describes the regression discontinuity design, an alternative to randomised experiments that allow researchers to allocate treatment to the subjects who are most in need while still being able to make strong causal inferences. The book ends with some closing thoughts in Chapter 8.

Space limitations prevented discussion of a few other important topics in causal inference: principal stratification methods, including complier average causal effects (CACE) models for analysing data from **broken experiments**, the Heckman selection model and fixed effects methods. However, the concepts covered in this book provide a solid foundation for understanding these methods. The concepts and methods that I chose to focus on have been the most useful for me and helpful for my students.

2

CONDITIONING

Chapter Overview

Conditioning is perhaps an unfamiliar word to many social scientists, but it describes a familiar concept. A conditional statistic is one that accounts for or incorporates external information, as opposed to a marginal statistic that does not (DeGroot & Schervish, 2002). For example, the marginal probability of rainy weather in some location might be 15%, but the probability of rain given that it is an overcast day in April might be 60%. The marginal mean (or expected) SAT score for college-bound high school graduates is around 1000, but the conditional mean SAT score for valedictorians might be 1500. Conditioning is the process by which this external information is incorporated into an analysis. According to modern theories of causal inference (Pearl, 2009; Rubin, 2005), inferring the existence and magnitude of causation between variables is possible even outside the context of a randomised experiment if the right set of conditioning decisions is made. This involves determining what extra information needs to be incorporated and how to incorporate it.

One way of thinking about conditioning is to imagine that the conditioned-on variables have been held constant. Pearl et al. (2016) described conditioning as a kind of filtering operation. For example, to condition on biological sex, an analysis may be performed only on females (or males). All conditioning methods remove variance in the outcome variable that is related to the conditioned-on variables. In the context of causal inference, this is necessary because some of the variance in the outcome variable often needs to be purged, as is the case if there is confounding, before valid causal inferences can be made. But *literally* holding a variable (or multiple variables) constant in an analysis is only one means of conditioning; this chapter describes several others. Some of these methods will be covered in more detail later in this book in the context of the causal inference techniques that rely on them.

The examples from this chapter will make use of a simulated data set which is based on the directed acyclic graph **(DAG)** or causal diagram displayed in Figure 2.1. Chapter 3 of this book will explain DAGs in much more detail. For now, it is sufficient to point out that the letters *A*, *B* and *C* represent variables and that the arrows represent causation. The variable *B* causes both *A* and *C*. The lack of an arrow connecting *A* with *C* indicates that *A* is not a cause of *C*.

A ⟵ - - - - - - - - - - B - - - - - - - - - - ⟶ C

Figure 2.1 Example of a DAG for the conditioning example data set

Note. DAG = directed acyclic graph.

This causal diagram depicts confounding. Variables *A* and *C* will be correlated due to their common cause (*B*) even though neither is causally related to the other. In other words, manipulating *A* would not cause *C* to change, nor would manipulating *C* cause *A* to change. This correlation will be detected by statistical analysis. When researchers are interested in causal inference, they would like for their statistical models to correctly estimate the magnitude of the causal connection between variables. However, any analysis of the relationship between A and C will yield an incorrect, non-zero estimate without conditioning on variable B.

To condition on B is to remove its effect on the outcome variable. This can be done in a literal sense by discarding all the data except for cases where B takes on a specific value, or nearly so. Or it can be done in a 'virtual' sense, such as by employing statistical adjustment (i.e. covariate adjustment). This chapter explores various conditioning techniques, provides some details of their implementation and illustrates how conditioning on confounders allows for causal effects to be estimated from non-experimental studies. Later chapters of this book, particularly Chapter 3 ('Directed Acyclic Graphs') and Chapter 5 ('Propensity Score Analysis'), provide additional context and elaboration of the ideas introduced here.

Simulated data set

I created a simulated data set from the scenario depicted in Figure 2.1. This data set, along with most of the others used in this book, is available for download on the book's companion website. The data set is formatted as *comma-separated values* (.csv) and can therefore be read by any statistical software, spreadsheet program or text editor.

This data set consists of $n = 4000$ observations for the variables A, B and C. The large sample size for this data set is not to suggest that one needs thousands of subjects in order to successfully apply these techniques, but rather to diminish the role of sampling error in the results that are presented; **bias** will be easily observable in these examples because the estimates will be several standard errors away from the true value of zero. But before examining the details, let us first consider the problems that conditioning methods are intended to solve: bias and inconsistency.

Bias and inconsistency

Sampling error is an important concept in statistical inference. The goal is to make inferences about the population quantities (parameters) from a sample, but samples never have exactly the same characteristics as the population because they are subsets of the population. Quantities calculated from samples (statistics) are rarely equal to their population counterparts, as they are approximations of those values. Researchers rarely care about the sample itself; rather, samples are studied as a vehicle for learning about the population. Statistics are estimators of population parameters (DeGroot & Schervish, 2002).

Consider a simple linear regression model of the following form:

$$Y_i = b_0 + b_1(X_i) + e_i \qquad (2.1)$$

This model in Equation (2.1) contains two statistics – the intercept term b_0 and the slope b_1 – which are calculated on the basis of a sample drawn from the population. We would not expect to observe precisely the same values of b_0 and b_1 if this model was fitted to another random sample drawn from the same population. The values would 'bounce around' due to sampling error. If this process could be repeated many times – draw a random sample from the population, fit the model to the sample and record the values of the statistics – the sampling distribution of each statistic would emerge.

The related concepts of bias and consistency are central to understanding the challenge of causal inference. They are commonly confused with one another, but describe different aspects of statistical performance. An unbiased estimator is one whose expected value is equal to its population parameter. Consistency describes the asymptotic behaviour of the estimator as the sample size grows (DeGroot & Schervish, 2002). A consistent estimator converges to the true value as the sample size approaches infinity; though it is asymptotically unbiased, it could exhibit bias in small samples. Therefore, unbiased and consistent estimators are generally preferred in statistics.

Let b represent the statistic and β its population counterpart, which in this case represents the true value of the causal effect. Equation (2.2) provides a formal definition of bias.

$E(b) = \beta$, an unbiased estimator

$\qquad (2.2)$

$E(b) \neq \beta$, a biased estimator

And Equation (2.3) provides a formal definition of consistency.

$\lim_{n \to \infty} b = \beta$, a consistent estimator $\qquad (2.3)$

$\lim_{n \to \infty} b \neq \beta$, an inconsistent estimator

The sampling distribution of the statistics of an unbiased estimator will be centred at the value population parameter regardless of the sample size. Some of the realised values of the statistic will be too small, others will be too large. But over the long run, across many repeated samples, these positive and negative errors will balance out exactly. The expected value (or mean) of the sampling distribution will be exactly equal to the population parameter.

When causal inference is the goal, the population parameter of interest is the causal effect of some variable on another. Unbiased estimation of causal effects means that the statistical estimates that are produced through our analysis are, on average, 'on target'. In real life, an infinite set of repeated studies cannot be performed; the study is done once using a single sample. As such, every estimated causal effect will be somewhat erroneous simply due to sampling error. But statistical theory allows us to place expectations on the degree of this 'wrongness'. And so long as the estimates are consistent, we can control the degree of this error – the difference between the estimated result and the true population causal effect – by adjusting the sample size of our study.

Figure 2.2 illustrates the behaviour of an unbiased and consistent estimator. The true population causal effect is zero, as shown by the horizontal reference line. Each dot is the b_1 statistic (e.g., the slope coefficient for predictor X) estimated from a simple regression model. Using simulation, I repeated this process thousands of times on many data sets of varying sizes sampled from the same population. As the sample size increases, the distribution of statistics is clearly converging to the true value, but the distribution is centred on the true value across the whole range of n. Unbiased and consistent estimators are the ideal, but if the estimator is at least consistent and asymptotically unbiased, the estimates can be made arbitrarily precise by making the sample size arbitrarily large.

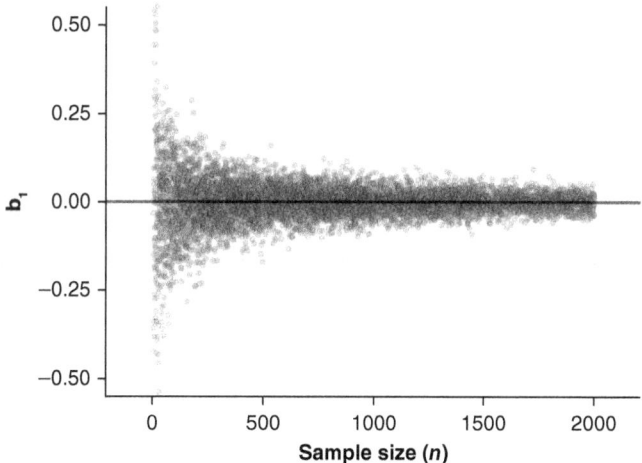

Figure 2.2 Illustration of an unbiased and consistent estimator. Sample size is on the x-axis. The estimated regression coefficient of A on C is on the y-axis. Each dot is a regression slope coefficient from a simulated study. The true value of the causal effect is depicted by the horizontal line

Figure 2.3 shows the behaviour of a biased but consistent estimator. As the sample size approaches infinity, both the bias and the sampling error approach zero, and estimates converge to the true population value. The familiar equations for calculating the sample variance and standard deviation (the ones that divide by n rather than $n - 1$) are biased but consistent estimators of their corresponding population parameters. Given that their assumptions are met, maximum likelihood estimators are consistent but not guaranteed to be unbiased, whereas least squares estimators are unbiased and consistent (Wooldridge, 2016). This type of estimator, which is biased for small sample sizes but approaches unbiasedness as the sample size goes up, is known as asymptotically unbiased.

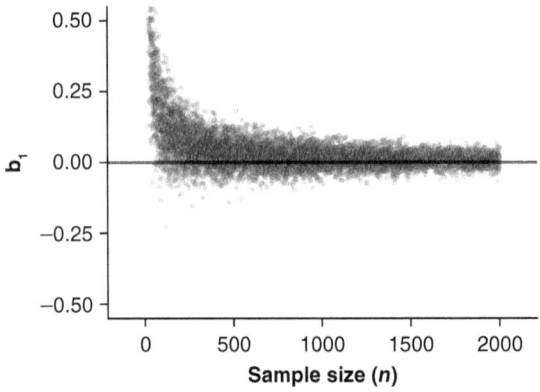

Figure 2.3 Illustration of a consistent estimator that is biased at small sample sizes. The estimator is convergent and asymptotically unbiased

An inconsistent estimator does not display this behaviour. Either the estimates do not converge at all or they converge to the wrong value. Figure 2.4 displays the behaviour of two inconsistent estimators. In the left panel, the sampling variance of the estimates becomes small as n gets large, but the estimates converge to the incorrect value of $\beta_1 = 0.5$. They are biased for all n. In the right panel, the estimates are unbiased because they are centred at the correct value, but they are inconsistent because they do not converge as a function of n. In either case, inconsistency means that the analyst cannot calculate the true uncertainty of an obtained estimate and cannot diminish it below some tolerable threshold by increasing the sample size.

The confounding illustrated in Figure 2.1 will cause biased and inconsistent estimates of the causal relationship between A and C. Since A and C are both caused by B (the confounder), any changes in B will manifest as contemporaneous changes in A and C, inducing a correlation between them even though neither variable causes the other.

The correlation induced between A and C could be eliminated if changes in B could be prevented. This is precisely what is accomplished by conditioning. Conditioning on B removes the spurious correlation between A and C, allowing for the unbiased

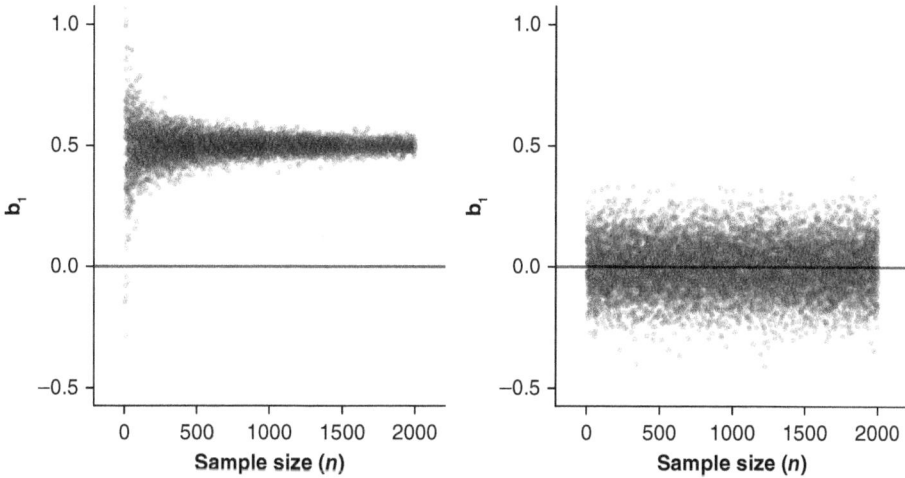

Figure 2.4 Two examples of inconsistent estimators. Left panel: The estimates converge to the wrong value as the sample grows. These estimates are both biased and inconsistent. Right panel: The estimates are unbiased but inconsistent; they do not converge to the true value

estimation of their actual (zero) causal connection. One can imagine conditioning as a type of operation that fixes or 'pins down' B to some specific value, preventing it from changing. According to the DAG in Figure 2.1, if B cannot change, it cannot transmit an association from A to C.

This text will often describe situations leading to biased estimates of causal effects. Unless otherwise indicated, these situations will also lead to inconsistent estimates as well. The problem of confounding is not one that can be corrected with a large n.

Obtaining a biased estimate of the causal effect

Our first analysis of the data will be a naive one that fails to condition on B. This will serve as a baseline against which the later results, which will implement various conditioning methods, can be compared. It illustrates how wrong the answer can be when confounding is not addressed either in the design of a study or in the data analysis. The result of this simple analysis will be badly biased and inconsistent.

I will arbitrarily choose C to be the response variable in this analysis, so this model estimates the causal effect $A \rightarrow C$. A simple regression model will be fit to the data of the following form:

$$C_i = b_0 + b_1(A_i) + e_i \tag{2.4}$$

where i indexes individuals and $e_1 \sim N(0,\sigma)$. In Equation (2.4), the slope coefficient b_1 estimates the unconditional effect of A on C. This unconditional model can be fit to the data in R using the lm() (*linear model*) function. Its verbatim output is given below.

```
##
## Call:
## lm(formula = C ~ A, data = data)
##
## Residuals:
##      Min      1Q  Median      3Q     Max
## -4.9519 -0.8142 -0.0061  0.8297  5.3013
##
## Coefficients:
##                Estimate Std. Error t value Pr(>|t|)
## (Intercept) -0.003883   0.019343  -0.201    0.841
## A            0.525887   0.013441  39.124   <2e-16 ***
## ---
## Signif. codes:  0 '***' 0.001 '**' 0.01 '*' 0.05 '.' 0.1 ' ' 1
##
## Residual standard error: 1.223 on 3998 degrees of freedom
## Multiple R-squared:  0.2769, Adjusted R-squared:  0.2767
## F-statistic:  1531 on 1 and 3998 DF,  p-value: < 2.2e-16
```

The results indicate that C is associated with A, $b = 0.526(0.013)$, 95% CI [0.500, 0.552], $p < .001$. The true value of zero is not captured within the confidence interval. However, the causal diagram describing the true data-generating process (Figure 2.1) does not contain an arrow connecting A with C. They are not causally connected, but are statistically related due to having the common cause B. This simple regression model has failed to estimate the true causal relationship between A and C.

Covariate adjustment

Perhaps the most familiar method of conditioning is to measure the variable and include it as a so-called control variable or covariate in a statistical model. In this way, the variable is virtually held constant as long as the model assumptions are met. This means of conditioning goes by many names, including *statistical control, regression adjustment* or *statistical adjustment*. These terms are synonymous.

In the context of linear regression, conditional effects are estimated by adding additional predictor variables to the model via a multiple regression model.

The regression slope coefficient reported for each predictor variable is interpreted as the expected effect of an isolated change on that variable on the outcome, holding all the other predictor variables constant. This 'holding constant' is a type of virtual conditioning (Fox, 1997).

For example, if the multiple regression model, $C_i = b_0 + b_1(A_i) + b_2(B_i) + e_i$, is fit to the data, the effect of variable A is computed conditioning on B. The slope b_1 represents the expected change in Y for an isolated 1-unit change in A given no change in B. Similarly, the slope b_2 is the expected change in C given an isolated change in B, unaccompanied by any change in A. Every regression coefficient in a linear regression model is estimated after conditioning on all the other predictor variables in the model. The results of fitting this model to the data are as follows:

```
##
## Call:
## lm(formula = C ~ A + B, data = data)
##
## Residuals:
##      Min      1Q  Median      3Q     Max
## -4.2932 -0.6621 -0.0084  0.6699  3.3759
##
## Coefficients:
##               Estimate Std. Error t value Pr(>|t|)
## (Intercept) -0.011776   0.015545  -0.758    0.449
## A            0.004139   0.015515   0.267    0.790
## B            1.009787   0.021555  46.847   <2e-16 ***
## ---
## Signif. codes:  0 '***' 0.001 '**' 0.01 '*' 0.05 '.' 0.1 ' ' 1
##
## Residual standard error: 0.983 on 3997 degrees of freedom
## Multiple R-squared:  0.5332, Adjusted R-squared:  0.5329
## F-statistic:  2283 on 2 and 3997 DF,  p-value: < 2.2e-16
```

Conditional on B, the effect of A on C is not significantly different from zero, $b = 0.004(0.016)$, 95% CI [−0.026, 0.035], $p = .79$. Note that the true value of zero is contained in the confidence interval. By conditioning on B, the model has correctly estimated the true causal effect of A on C.

Visualising covariate adjustment

Figure 2.5 illustrates the source of the biased result from the simple regression model, $C_i = b_0 + b_1(A_i) + e_i$. The issue is that change in A does not occur in isolation, but is paired with expected change in B. This is because of the $B \rightarrow A$ path in the causal diagram; *B* causes *A*, so their changes must co-occur. As *B* increases, *A* increases with it. But because of the path *B→A*, *C* increases as well. Thus, variation in *A* tends to co-occur with variation in *C*. The regression model properly describes this associational but non-causal feature of the data. To apprehend the correct causal relationship between *A* and *C*, change in *B* must be prevented. In other words, one must condition on *B*. This isolates the expected change in *C* if *A*, and only *A*, were manipulated. This is the effect estimated for *A* in the multiple regression model when *B* is included as a covariate (Fox, 1997).

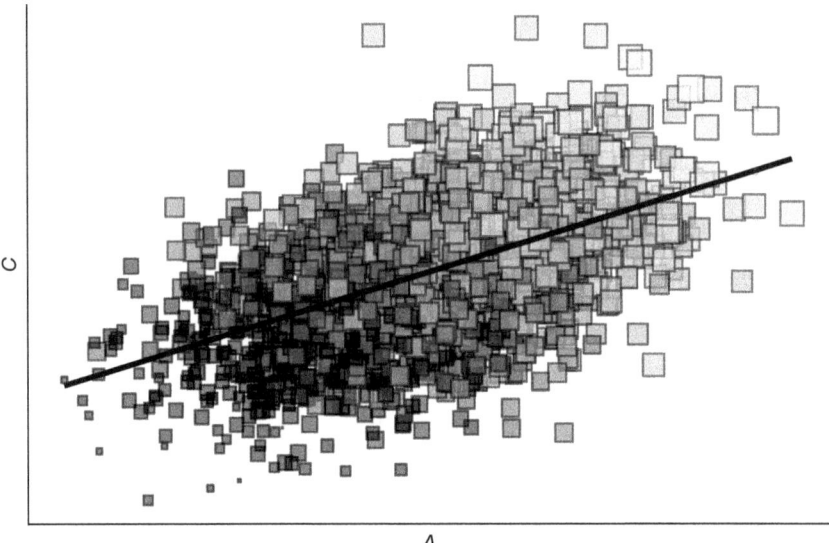

Figure 2.5 Simple regression of *C* on *A*. Values of *B* are depicted by point shading and size, with bigger, lighter points higher on *B*. Because *A* causes *B*, cases with high values of *A* also tend to have high values of *B*. And because *B* causes *A*, cases with high values of *B* tend to have high values of *C*. This creates a spurious relationship between *A* and *C* that is portrayed by the regression line

Figure 2.6 shows the data plotted against this multiple regression model's response surface in three dimensions from four viewing angles. The response variable, *C*, is on the *z*-axis of the plot, with the predictors *A* and *B* on the *x*- and *y*-axes. The regression model projects the data into an orthogonal space within which the effect of each predictor variable can be isolated. The top left panel views the plot from 'above', that is, along the *z*-axis with respect to *C*. This panel shows a scatter plot of *A* versus *B*, which are moderately correlated. This correlation is due to their causal connection; *B* causes *A* and *C* according to Figure 2.1.

The bottom left panel of the figure displays the plot rotated such that the regression plane is viewed with respect to *B*. The positively sloped regression line (when viewed from this perspective) indicates that isolated change in *B* is associated with change in the response variable, C. This notion of 'isolated change' is exactly what is accomplished by covariate adjustment. Clearly A and B are correlated. In the sample, change in A is frequently paired with nearly proportional change in B. But from the orientation depicted in this panel, such correlation has been rendered irrelevant. As the value of B increases, so does the value of A, but this merely moves the points further away from the viewer's perspective as viewed from this angle and has no effect whatsoever on the slope of the relationship between B and C.

In the lower right panel, the plot has been rotated such that it is viewed with respect to A. The regression solution is flat from this angle, depicting the null relationship between A and C conditional on B. Once more, the fact that A is not separable from B in the sample (e.g. in reality) has been rendered irrelevant. The contemporaneous

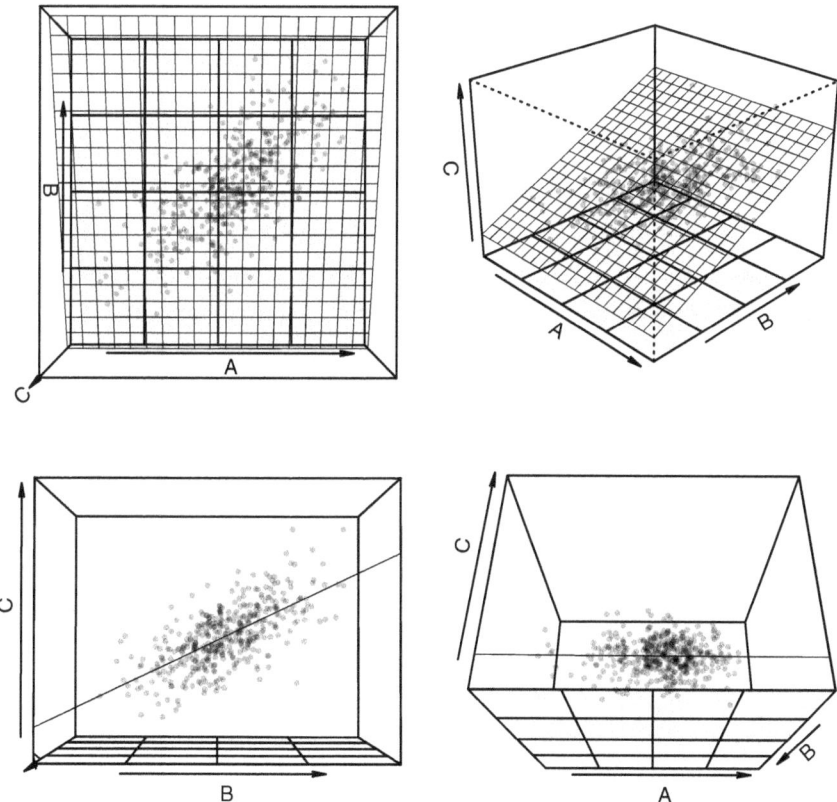

Figure 2.6 Conditioning via covariate adjustment with linear regression. Upper left: 3D scatter plot of A (x-axis) and B (y-axis) versus C (z-axis) viewed from above (with respect to C). Upper right: Rotation to show the points plotted against the regression surface. Lower left: Rotation to view the regression surface with respect to B. Variation on A is invisible and irrelevant from this perspective, as it occurs along the line of sight of the observer. Lower right: Rotation to view the regression surface with respect to A. Variation on B is invisible and irrelevant from this perspective

change in *B* accompanying change in *A* (due to their correlation, which reflects the causal link *B*→*A*) has been projected into a dimension that cannot disturb the estimated relationship between *A* and *C*; that dimension being away from the viewer's eye, into the page.

Linear regression and related models perform the magic of conditioning by projecting correlated data into an orthogonal space in which each dimension represents a variable. In this way, each variable's isolated effect can be determined. When viewed from the perspective of each predictor, the regression solution renders the variability on the other predictors irrelevant. This process becomes impossible to visualise when there are more than two predictors, but the concept of conditioning via statistical control generalises to any number of predictors (James et al., 2013).

Covariate adjustment depends on strong assumptions

The quality of estimates generated from linear models for both unconditional and conditional effects depends on the degree to which the model assumptions are met. Unmodelled non-linearity or measurement error in the predictor variables jeopardises the model performance by biasing the regression coefficients. Further, because of the nature of conditional estimation, specification errors or measurement errors involving one predictor variable can bias the estimated coefficients for *all* of the model's predictors. Problems involving one variable can affect other variables' estimates and contaminate the whole model (Fox, 1997). For this reason, exposure to problems related to assumption violations scales with the number of predictors. Further, a much larger set of assumptions must be satisfied for the standard errors (and therefore confidence intervals and statistical hypothesis test results) to be correct. More details about these issues may be found in Volume 8 (*Linear Regression: An Introduction to Statistical Models*)

These assumptions are met in this simulated data set, but may not be satisfied in a real research context. Conditioning via covariate adjustment is probably the most convenient and simplest to implement of all the conditioning methods, but it is also the most assumption laden and therefore the least likely to work correctly in practice (Ho et al., 2007).

Sample selection

Sample selection, which could also be called data rejection or restriction, is the most literal type of conditioning. Researchers condition via sample selection when they ensure that their sample contains limited variation with respect to some variable. If a researcher wishes to condition on, for example, temperature, she could ensure that the lab temperature is always maintained at the same level. The word 'controlled' in the phrase 'controlled experiment' refers to this type of conditioning, in which all factors that could conceivably affect the outcome variable are held constant across experimental conditions. To condition on age, gender or race via sample selection, researchers could collect data only from subjects meeting particular inclusion criteria, discarding the rest.

To illustrate the effect of conditioning on B via sample selection, I have selected a subset of the data in which the variable B contains only a narrow range of values, $-0.1 < B < 0.1$. The new data set has been reduced from $n = 4000$ to $n = 320$. Any analysis based on the reduced data set implicitly conditions on B. The results of fitting the simple regression model from equation 2.4 to this data are as follows:

```
## 
## Call:
## lm(formula = C ~ A, data = data.subset)
## 
## Residuals:
##      Min       1Q  Median       3Q      Max
## -2.67562 -0.68435 0.04741 0.65987 2.36123
## 
## Coefficients:
##             Estimate Std. Error t value Pr(>|t|)
## (Intercept)  0.01200    0.05424   0.221    0.825
## A            0.03019    0.05238   0.577    0.565
## 
## Residual standard error: 0.9698 on 318 degrees of freedom
## Multiple R-squared:  0.001044,   Adjusted R-squared:  -0.002097
## F-statistic: 0.3324 on 1 and 318 DF,  p-value: 0.5647
```

The results reveal that A is unrelated to C, $b = 0.03(0.052)$, 95% CI [$-0.072, 0.133$], $p = .565$. The true value of the causal effect (0) is contained in the confidence interval.

One problem with conditioning by sample selection is that it can harm the external validity of the study. For example, nearly all behavioural neuroscience research is based on male rats. Researchers typically exclude female rats from experiments due to concerns about behavioural changes during oestrus, though those concerns may be overblown (Dayton et al., 2016). The generalisability of these findings to female rats is therefore questionable. Further, a large proportion of psychology research is done on college students. In addition to harming external validity, conditioning via sample selection can involve discarding huge amounts of data, particularly when conditioning on a continuous variable or a low-incidence categorical variable. Imagine how many potential participants would be turned away if a full-scale IQ (intelligence quotient) score of exactly 100 was required to enroll in a study.

The bias–variance trade-off

The variance of a statistical estimator is the magnitude of its sampling variability or standard error. The larger this is, the less **precision** the estimate will have. High-variance estimates provide less information about the quantity being estimated. Low variance is precision, which manifests as narrow confidence intervals and high statistical power for hypothesis tests. Unfortunately, procedures that reduce bias often increase variance. Reducing bias comes at a cost (James et al., 2013).

This trade-off is quite intuitive in the context of sample selection as a conditioning method. Estimation bias of the $A{\rightarrow}C$ causal effect is a function of B's variability; eliminating this variability completely would result in a completely unbiased estimate of the causal effect. If B is a continuous variable, as it is in this example, this would also entail discarding nearly all of the sample. Some variability on B must be allowed to make the analysis possible. The wider the selection bandwidth becomes, the more bias is admitted into the estimation process, but the larger the effective sample size becomes.

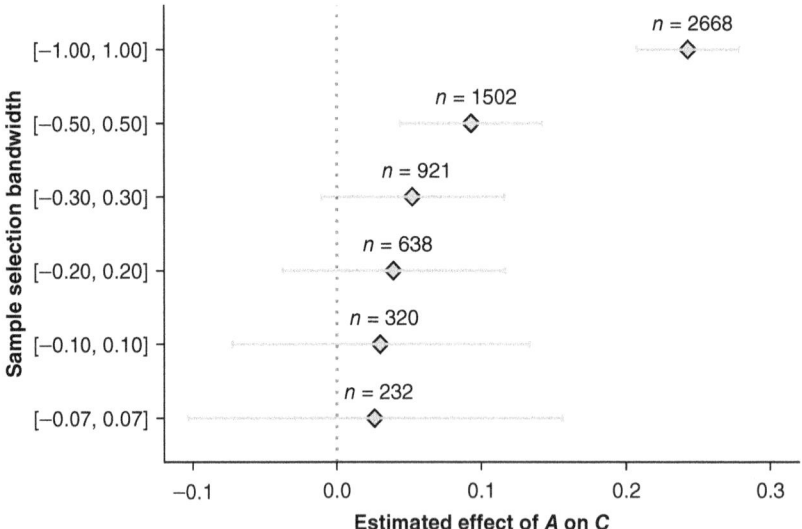

Figure 2.7   The bias–variance trade-off. The figure displays point estimates and 95% confidence intervals by sample selection bandwidth. The dotted vertical line indicates the true value of the causal effect. The values on the x-axis represent the sample selection bandwidth, which is the range of allowable values on variable B that were retained in the analysis. Wider bandwidths retain a larger sample and therefore exhibit less sampling variance (narrower confidence intervals), but are more biased

Because standard errors are inversely proportional to the sample size, larger samples mean reduced estimation variance (DeGroot & Schervish, 2002). Figure 2.7 shows the consequences of different choices for the sample selection bandwidth on both bias and variance. Bias is represented as the distance from the point estimate to the true value of A's effect on C (zero), variance by the width of the confidence interval.

The bias–variance trade-off is an unavoidable feature of all non-parametric or non-model-based conditioning techniques. It is less applicable to covariate adjustment because statistical models replace the data itself with a parametric function of the data which is assumed to be correct and complete. However, as we have seen, covariate adjustment can produce biased results when the assumptions of the statistical model are violated.

Subclassification

In subclassification, which is also known as stratification, the sample is divided into subgroups either on the basis of a categorical variable or by dividing a continuous variable into discrete groups (Pourhoseingholi et al., 2012). For example, researchers wishing to condition on IQ might divide the sample into IQ ranges, such as 90 to 100, 101 to 110, 111 to 120 and 121 to 130. Statistical quantities of interest, such as the association between a focal variable of interest and the outcome variable, are computed separately for each stratum. The variation on IQ is therefore reduced within strata. The marginal treatment effect (i.e. the overall treatment effect across all the strata) can be estimated by calculating a weighted average of all the strata-specific estimates, allowing the more populated strata to carry more weight in the final estimate.

I used R's cut() function to divide B into a set of ten strata. The table() function counts the number of cases falling into each stratum. The output of this function is given below:

```
##
##   1   2   3   4   5   6   7   8   9  10
## 41 106 302 617 971 852 652 311 127  21
```

Estimating the regression coefficients for the effect of A on C within each stratum can be accomplished in many different ways. I computed them using a regression model, which is written in matrix form for compactness.

$$C_i = \alpha \left(\textbf{Strat}_i \right) + \gamma \left(A_i \times \textbf{Strat}_i \right) + e_i \tag{2.5}$$

where $\textbf{Strat}_i$ is the ith subject's vector of 10 dummy-coded stratum indicator variables. Note that the model does not contain an intercept, allowing all 10 stratum dummies to be included in the model. The vector α is a set of 10 stratum-specific intercepts, and γ is a vector of 10 stratum-specific slope coefficients for the effect of A. Weights were incorporated into the analysis (via functions from the R package survey) to account for the varying stratum sample sizes in calculating the standard errors of the γ parameters.

Table 2.1 provides the regression results for the effect of A on C within strata. These are the γ parameters from Equation (2.5), as well as their standard errors and p-values. None of these within-strata regression slopes was statistically significant at the .05 level. Remember that the risk of making at least one Type I error increases with the number of statistical tests that are performed. Given a set of 10 statistical tests, it should be unsurprising if one or two are statistically significant even when the null hypothesis is true.

Table 2.1 Stratum-specific regression results

Stratum	n	Range	b (SE)	p
1	41	$(-\infty, -2.50]$	0.030 (0.142)	.835
2	106	$[-2.50, -1.875]$	0.013 (0.102)	.899
3	302	$[-1.875, -1.25]$	0.045 (0.048)	.348
4	617	$[-1.25, -0.625]$	0.025 (0.038)	.508
5	971	$[-0.625, 0.00]$	0.051 (0.030)	.091
6	852	$[0.00, 0.625]$	0.054 (0.034)	.118
7	652	$[0.625, 1.25]$	0.045 (0.039)	.243
8	311	$[1.25, 1.875]$	−0.010 (0.064)	.877
9	127	$[1.875, 2.50]$	−0.022 (0.094)	.814
10	21	$[2.50, \infty)$	−0.029 (0.126)	.817

The stratum-specific test can be statistically significant for another reason besides Type I errors. When the conditioning variable is continuous, as is the case in this example, there is always some residual variation on that variable within strata. Stratifying on a continuous variable cannot remove all of its variance. If too few strata are created, the degree of conditioning achieved may be insufficient to permit valid inference. But too many strata increase the standard error of the overall estimate, widening the confidence interval and reducing its statistical power. As is the rule in statistics, subclassification involves a bias–variance trade-off. Increasing the number of strata reduces bias but increases the variance of the estimate. Generally speaking, larger sample sizes permit researchers to afford more strata. The problem of residual variation within strata does not occur when the stratification variable is categorical and measured without error.

The stratum-specific tests are uninteresting; they need to be *pooled* or combined into a single result. I pooled these results by computing a custom hypothesis test using the glht() function from the multcomp package, applying coefficients of $\left(\frac{1}{10}\right)$ to each of the γ parameters. This computes the pooled estimate along with its standard error and hypothesis test result. The coefficients should all be $\frac{1}{10}$ instead of proportional to the stratum sample sizes because the weights were incorporated into the analysis at the previous step. The output from this function is shown below.

```
##
## Simultaneous Tests for General Linear Hypotheses
##
## Fit: svyglm(formula = C ~ 0 + stratF + A:stratF, design = design.strat)
##
## Linear Hypotheses:
##          Estimate Std. Error z value Pr(>|z|)
## 1 == 0 0.02018    0.02582    0.782    0.434
## (Adjusted p values reported -- none method)
```

This, the slope coefficient for the effect of A on C, aggregated over all the strata, is b = 0.02(0.026), 95% CI [−0.030, 0.071], p = .434. After conditioning on B via subclassification, the estimated relationship between A and C is extremely weak and not significantly different from zero.

Unlike covariate adjustment, the validity of subclassification as a conditioning tool does not depend on any assumptions regarding the functional form of the relationship between the outcome variable and the conditioning variable. This is its major advantage. Disadvantages of subclassification are that one can only condition on a *very* small set of variables (typically one), that it is impossible to remove all of the variance of continuous conditioning variables due to residual variation with strata, and the bias–variance trade-off. And like all conditioning methods, complete conditioning requires perfect measurement of the conditioning variable.

Matching

In the simplest version of matching, the treatment variable is dichotomous. Each treated subject is matched with a control subject with similar or identical values of the conditioning variable. The treatment effect is computed for each matched pair, and causal effect by averaging over them. In reality, there are many other matching methods, including matching with and without replacement, one-to-many matching, greedy matching, full matching, nearest neighbour matching, to list only a few of the diverse matching algorithms described in the literature (e.g. Stuart, 2010).

Many of these are described in much more detail in Chapter 5 of this book. However, they all have in common the notion of reducing or eliminating variability on the matching variable(s).

Matching requires that the predictor variable of interest – the *focal variable* – be categorical. The focal variable (*A*) in the example data set is continuous. In order to provide a demonstration of conditioning via matching, I have created a dichotomised version of the A variable called *Acat* ('A categorical'), which will be set to one when $A \geq 0$ and zero otherwise. (*Note:* Categorising continuous variables as a general analytic strategy is rarely warranted; I am doing it here only to make it possible to demonstrate matching as a conditioning strategy; see Maxwell & Delaney, 1993.)

Dichotomising the *A* variable does not substantially change the pattern of relationships between variables. A simple regression analysis estimating the effect of *Acat* on *C* continues to report a strong relationship between these variables. To demonstrate this fact, the output below shows the results of fitting a simple regression model of form $C_i = b_0 + b_1 (Acat_i) + e_i$ to the data.

```
##
## Call:
## lm(formula = C ~ Acat, data = data)
##
## Residuals:
##     Min      1Q  Median      3Q     Max
## -5.0721 -0.8623  0.0077  0.8811  5.5747
##
## Coefficients:
##               Estimate Std. Error t value Pr(>|t|)
## (Intercept) -0.61072    0.02924  -20.89   <2e-16 ***
## Acat         1.19985    0.04135   29.02   <2e-16 ***
## ---
## Signif. codes:  0 '***' 0.001 '**' 0.01 '*' 0.05 '.' 0.1 ' ' 1
##
## Residual standard error: 1.307 on 3998 degrees of freedom
## Multiple R-squared: 0.174,  Adjusted R-squared: 0.1738
## F-statistic: 842.2 on 1 and 3998 DF,  p-value: < 2.2e-16
```

Of course, the regression model is not *wrong* as a mere description of the data, as there is strong statistical dependence between *Acat* and *C* by virtue of their common cause. But as an estimate of the causal effect of *Acat* on *C*, which we know from the causal diagram is actually zero, this result is severely biased and misleading (Berk, 2004). Correctly estimating this causal relationship requires conditioning on *B*.

The Matching package (Sekhon, 2019) is one of several in R that implements matching. I used the Match() function from that package to perform one-to-one matching without replacement with a caliper of 0.1 *SD* (see Chapter 5 for more details about matching, where these terms are explained). A **caliper** value specifies the maximum tolerable degree of dissimilarity between matched cases; cases that cannot be matched within this tolerance are discarded (Lunt, 2013). Figure 2.8 displays a scatter plot of values of *B* for *Acat* = 0 and *Acat* = 1 after matching. All of the points would be found on the diagonal reference line under perfect matching. The dotted diagonal lines represent the caliper value of 0.1 that I specified. Note that this plot is quite 'zoomed in'. Figure 2.8 indicates that *B* was matched quite closely across groups of *Acat*.

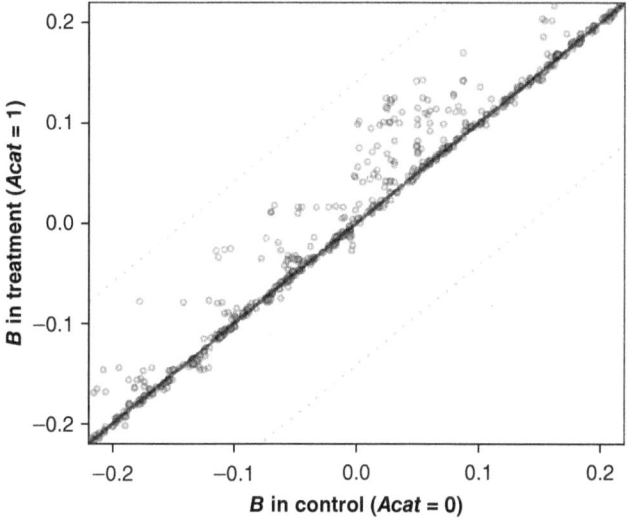

Figure 2.8 Conditioning via matching: Scatter plot of $B_{treatment}$ versus $B_{control}$ after matching. Each point represents one matched pair. Dotted lines indicate caliper boundaries

Table 2.2 shows descriptive statistics before and after matching. This provides another method for checking the effectiveness of matching as a means of conditioning on *B*. Note that the summary statistics of *B* are quite similar across levels of *Acat* after matching. And indeed, the results confirm that matching worked quite well as a means of conditioning on *B*.

Table 2.2 Descriptive statistics before and after matching on *B*

	Mean Acat = 1	Mean Acat = 0	SD Acat = 1	SD Acat = 0
Before matching	0.596	−0.593	0.850	0.847
After matching	−0.016	−0.022	0.608	0.614

Having confirmed that the matching process has effectively conditioned on *B*, running the `summary()` function on the `Match()` function's output produces the estimated effect of *Acat* on *C*, conditional on *B*. This output is given below.

```
##
## Estimate...   -0.0030122
## SE........   0.031661
## T-stat.....   -0.095138
## p.val......   0.92421
##
## Original number of observations.............  4000
## Original number of treated obs..............  2000
## Matched number of observations..............  1970
## Matched number of observations  (unweighted).  1970
##
## Caliper (SDs)...............................  0.1
## Number of obs dropped by 'exact' or 'caliper'  2030
```

The matched analysis reveals that this relationship is tiny in magnitude and not statistically distinguishable from zero, $b = -0.003(0.032)$, 95% CI [−0.065, 0.059], $p = .924$.

Like subclassification, matching involves a bias–variance trade-off. Smaller calipers force increasingly exact matches, reducing bias. But this has the unavoidable side effect of excluding larger fractions of the sample, increasing variance – which implies reduced estimation precision and statistical power. Large calipers retain more of the data but allow for larger amounts of residual variation on the matching variable(s) within matched sets, increasing bias. The output above reveals that 2030 of the 4000 cases in the sample were not matched. This discarding of data increases the variance of the estimate but also reduces its bias. In this example, the remaining 1970 cases are sufficient to produce quite a narrow standard error of 0.032, and therefore a 95% CI with a width of 0.063.

The primary virtues of matching as a conditioning strategy are that no functional form is assumed regarding the relationship between the response variable and the conditioning variable, and that diagnostics regarding the efficacy of matching are readily available. Although there are numerous options to choose from when implementing matching (e.g. with or without replacement? one-to-one or one-to-many?), the best set of choices produces the optimal balance on the conditioning variables while retaining as much of the data as possible. Furthermore, since matching and the estimation of causal effects occur in separate steps (unlike, e.g., covariate adjustment) researchers are able to try out different matching strategies in search of optimality without seeing how those choices affect the final result of the study. This helps to prevent researchers from unconscious *p*-hacking by selecting the matching method that produces a favoured result (Wicherts et al., 2016).

Weighting

Weighting is yet another conditioning method. Like matching, weighting can only be used for conditioning when the focal variable is categorical (and preferably dichotomous). The probability distribution of the variable to be conditioned on is estimated within each level of treatment. For example, imagine that the treatment being investigated is a GRE (Graduate Record Examination) preparation course, and that the conditioning variable is academic achievement. The 'treatment group' would consist of those students who enrol in the course; the 'control group' those who do not. The left panel of Figure 2.9 shows that the distribution of academic achievement varies across treatment groups, with the enrollees having a higher mean achievement level. If academic ability influences students' decisions to enrol in the prep course, then correctly estimating the causal effect requires conditioning on academic achievement. Otherwise, group differences observed on GRE scores could simply reflect pre-existing differences in achievement rather than any true effect of the GRE prep course. The right panel of Figure 2.9 shows the distribution of academic ability by group after weighting. The weights equalise the distribution of academic ability across groups (Austin & Stuart, 2015).

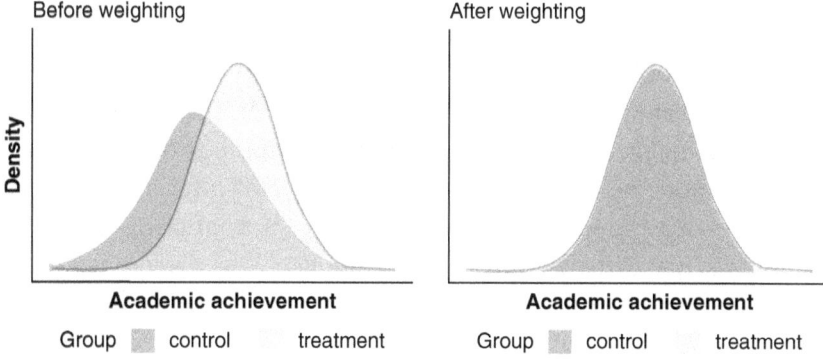

Figure 2.9 Conditioning via weighting. Left panel: Distribution of academic ability by treatment group before weighting. The target distribution is outlined. Right panel: Distribution of academic ability by group after weighting. The distribution of academic ability in the control group has been altered by the weighting to resemble its distribution in the treatment group

Computing the weights

The weights are computed based on the densities of each sample value observation of the conditioning variable in each level of the focal variable. In the sample data set, the conditioning variable is B and the focal variable is *Acat*, which has two levels, which I will refer to as 'treatment' and 'control' without intending to imply inter-vention or random assignment. A density is the likelihood of observing a particular value of a random variable under a given probability density function. For example, under a standard normal distribution, with a mean of zero and a standard deviation of one, an observation of $x = 0.1$ has a much higher density than an observation of $x = 3.2$, because the first point is near the mean of the distribution, while the second is located out in the tail.

When the probability distribution of a variable is known, the density of any obser-vation can be calculated using equations derived from statistical theory that are com-monly implemented in statistical software (DeGroot & Schervish, 2002). Probability densities under a normal distribution are calculated using R's dnorm() function. Probability densities have only a relative interpretation, not an absolute one, because the probability of observing *any* specific value under a continuous distribution is zero. But some values are much more likely to be observed than others, and that is what their densities represent.

In reality, the distribution of the conditioning variable in each group rarely fol-lows a textbook probability distribution. It is unknown, but it can often be esti-mated quite reliably from the data using **kernel density estimation**. Kernel density estimation is a non-parametric technique for estimating the empirical prob-ability density function of a random variable (Silverman, 1986). Most readers will be familiar with the concept of a histogram, which is constructed by dividing the variable into a set of bins, counting the number of cases that fall within each bin (the frequencies), and plotting them via a bar graph. In practice, frequencies are often converted to densities by dividing by the total number of cases; in this way, the histogram plots the *proportion* of cases in each 'bin' instead of the *number* of cases. This prevents the scale of the y-axis of the plot from being affected by the sample size. A kernel density estimate can be thought of as a smoothed histogram. The 'kernel' in kernel density estimation refers to the mathematical function for implementing the smoothing.

The density() function in R calculates density estimates. The approx() func-tion is required to perform linear interpolation on the values returned by the density() function so that the densities are calculated for the specific values of the conditioning variable in the data rather than for a set of representative points spanning its range.

The next step, after calculating the estimated density of each observed value of *B* under each level of *Acat*, is to calculate the weights. There are three choices. One can apply weights to the control group (to make it resemble the treatment group), to the treatment group (to make it resemble control) or to both groups to make them both resemble the marginal distribution of the conditioning variable (Guo & Fraser, 2015).

1   The control cases ($Acat = 0$) can be weighted to give them a similar distribution on B as the treated cases. In this case, a weight of 1 would be assigned to all the treated cases, and weight of $\dfrac{\text{density}(B_i)\,|\,Acat = 1}{\text{density}(B_i)\,|\,Acat = 0}$ to all the control cases.

2   The treated cases ($Acat = 0$) can be weighted to give them a similar distribution on B as the control cases. The control cases would all receive a weight of 1.0, and weight of $\dfrac{\text{density}(B_i)\,|\,Acat = 0}{\text{density}(B_i)\,|\,Acat = 1}$ would be assigned to all the treated cases.

3   Weights can be applied to both groups to push each one to the marginal density. The treated cases would receive a weight of $\dfrac{\text{density}(B_i)}{\text{density}(B_i)\,|\,Acat = 1}$ and the control cases a weight of $\dfrac{\text{density}(B_i)}{\text{density}(B_i)\,|\,Acat = 0}$.

The underlying research question provides the rationale for making this choice. In a nutshell, strategy 1 is used to understand the causal effect of the focal variable on the subjects in the treatment group. Strategy 2 is used to understand the causal effect of the focal variable on the subjects in the control group. (In other words, *if* those subjects had been in treatment condition, how would their outcomes have changed?) And strategy 3 is used when the general or marginal effect of the treatment is of interest. The resulting causal estimates under these three strategies can differ due to the possibility that the conditioning variable B moderates the causal effect. This topic will be explored much more thoroughly in the later chapters on 'Rubin's Causal Model and the Propensity Score' (Chapter 4) and 'Propensity Score Analysis' (Chapter 5).

Table 2.3 displays the value of *Acat* (e.g. treatment status), the densities under $Acat = 0$ and $Acat = 1$ and the calculated weight for a selection of example cases.

Table 2.3 Weights versus density under $Acat = 0$ and $Acat = 1$ for treatment and control cases

Density under Acat = 0	Density under Acat = 1	Weight	Treatment
0.220	0.428	1.945	0
0.283	0.413	1.000	1
0.494	0.213	0.432	0
0.493	0.290	0.589	0
0.233	0.424	1.000	1

The first case has an *Acat* value of 0, so it was in the control group. Comparing its densities reveals that it is more likely to have been observed in the treatment group (density = 0.428) than in the control group (density = 0.22). In other words, it is more similar to a treatment case than to a control case with respect to the condition-ing variable B. Therefore, this case receives a weight of $\dfrac{0.428}{0.22} = 1.945$. Because this weight is larger than one, the first case has been *upweighted* in the analysis.

The second case was in the treatment group ($Acat = 1$), so it gets a weight of one. The third case is a control case. Its density is lower under the treatment distribution (density = 0.213) than under the control distribution (density = 0.494), so it receives a weight of less than one – it is *downweighted* in the analysis. The same thing occurred for the fourth case, though it is not downweighted quite as much. The fifth case is in the treatment group and receives a weight of one.

The svydesign() and svyglm() functions from the survey package can be used to perform the analysis once the weights have been calculated. The results of this analysis are reported below.

```
## 
## Call:
## svyglm(formula = C ~ Acat, design = design)
## 
## Survey design:
## svydesign(ids = ~1, weights = ~weights, data = data.complete)
## 
## Coefficients:
##               Estimate Std. Error t value Pr(>|t|)
## (Intercept)   0.54462    0.09281   5.868  4.76e-09 ***
## Acat          0.04451    0.09731   0.457   0.647
## ---
## Signif. codes:  0 '***' 0.001 '**' 0.01 '*' 0.05 '.' 0.1 ' ' 1
## 
## (Dispersion parameter for gaussian family taken to be 1.756351)
## 
## Number of Fisher Scoring iterations: 2
```

Note that the variable B was not included as a control variable in this analysis. The conditioning on B is accomplished via the weights. The analysis correctly reveals no association between $Acat$ and C, $b = 0.045(0.097)$, 95% CI [–0.146, 0.235], $p = .647$.

The problem of measurement

Complete removal of confounding bias via conditioning requires perfect measurement of the offending variables (Westfall & Yarkoni, 2016). This presents a paradox in that perfect measurement is impossible. In the social sciences, achieving even a vaguely acceptable level of measurement quality can be quite difficult. In reality, some confounding bias will persist even after conditioning. The severity of this residual confounding depends on the quality of measurement. Distressingly, though, it is clear that some of the typical levels of measurement reliability observed in the social sciences result in significant leftover confounding that cannot be removed.

It is important to note that the challenge of measurement is really about validity, but this section focuses on reliability (rather than validity) for two reasons. First, reliability is a prerequisite for validity. In fact, the reliability coefficient sets an upper bound on possible validity coefficients (Crocker & Algina, 1986). Second, there is an elaborated mathematical framework for discussing **reliability** that will be used to inform the simulations and analytic results that will be presented. While it is possible to have high reliability with low validity (e.g. by precisely measuring the wrong construct), in the discussion to follow, I will assume that the only cause of invalidity is noise in the measured values – in other words, imperfect reliability.

Classical test theory model for measurement error

The *classical test theory* (CTT) model is a basic framework for understanding measurement. While newer theories of measurement have somewhat supplanted CTT in applied measurement contexts, CTT remains immensely valuable for elucidating the principles of measurement. CTT is similar to Newtonian physics. While somewhat oversimplified, it works well enough in most situations.

According to the CTT mode, all observed scores (X) are composed of true score (T) and measurement error (E). Observed scores are the values obtained via an act of measurement. Crocker and Algina (1986) gave the following equations defining the CTT model.

$$X = T + E \tag{2.6}$$

The **true score** (T) is the actual quantity that we wish to observe: a person's true level of depression, academic achievement, voting intention or intelligence. Theories are statements about relations between true scores. Unfortunately, true scores are impossible to observe. All measurements are contaminated with some degree of **measurement error** (E). The values that can be obtained are **observed scores** (X). By Equation (2.6), these are envisioned as a sum of true score and error. Though imperfect, observed scores are useful to the extent that the measurement error component is small.

The CTT model holds that the variance of the observed scores, X, is the sum of the true score variance plus the error variance.

$$\text{var}(X) = \text{var}(T) + \text{var}(E) \tag{2.7}$$

These can be added because of the assumption that the measurement errors are independent of the true scores, $T \perp E$ implying that they have a covariance (and correlation) of zero. Therefore there is no covariance term to deal with when calculating the total variance (DeGroot & Schervish, 2002).

According to the CTT model, measurement errors are envisioned as disturbances drawn at random from a normal distribution with a mean of zero. Therefore measurement error does not systematically bias measurement by shifting the mean of the observed scores away from the mean of the observed scores. It is simply Gaussian noise – additional, meaningless and unwanted variability in measurement.

Reliability

The definition of reliability in the CTT model is the ratio of true score variance to total variance.

$$\rho_{xx} = \frac{\text{var}(T)}{\text{var}(T) + \text{var}(E)} \tag{2.8}$$

In other words, the *reliability* (ρ_{xx}) of X is essentially an r^2 statistic, as it represents the proportion of variance in the observed score (X) that is true score (T). The interpretation is very straightforward (Crocker & Algina, 1986).

Therefore, the correlation between the true score and the observed score is the square root of the reliability. This fact makes it relatively easy to simulate data with measurement errors. Figure 2.10 displays the results of conditioning on three versions of the B variable, each contaminated with various degrees of measurement error manifesting as varying levels of reliability. For context, the upper left panel of Figure 2.10 illustrates the result of omitting B from the model entirely. Similar to Figures 2.2 and 2.4, each point represents an estimated value of the regression coefficient in a simulated study. As the sample size increases, the estimates should converge to the true causal effect for an unbiased estimator. But this does not occur when the conditioning variable is measured with error. The estimates in the upper right, lower left and lower right panels are biased in spite of confounder B's inclusion in the model. The degree of bias is a consequence of the reliability with which variable B has been measured.

This issue of measurement error has a deleterious effect on the ability of all conditioning methods to produce unbiased estimates of causal effects. One may implement the sample selection method using a very narrow bandwidth, but even when cases are highly similar on the observed values of the conditioning variable, they may remain relatively dissimilar on the unobservable true values of that variable. In subclassification, measurement error disturbs the stratum assignment of cases.

While we might like to stratify on the true score, we can only stratify on the noisy observed scores, and this increases the dissimilarity of cases within strata.

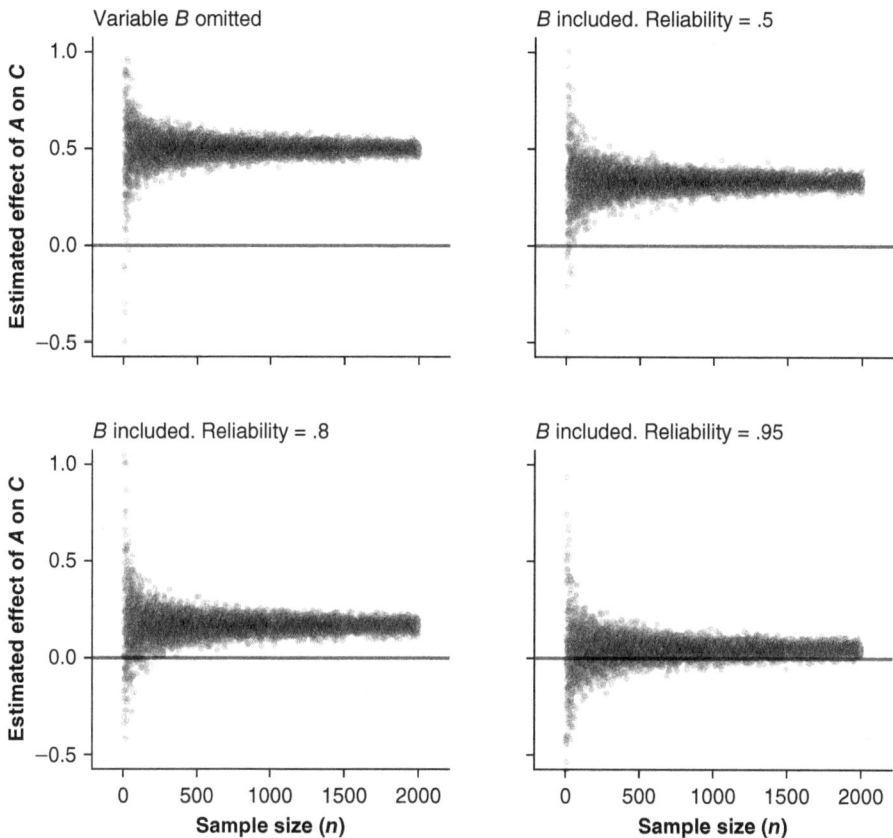

Figure 2.10 Biased and inconsistent estimates when conditioning on an imperfectly measured confounder. Sample size is on the x-axis. The estimated regression coefficient of A on C is on the y-axis. Each dot is a regression coefficient from a simulated study. Bias occurs when the estimates do not converge to their true value (zero) as the sample size increases. Upper left: Confounder variable omitted from the model. Upper right: Conditioning on a poorly measured confounder. Lower left: Conditioning on an 'adequately' measured confounder. Lower right: Conditioning on a superbly measured confounder

Completely unbiased estimation of causal effects is an unrealisable ideal in most situations outside of strictly controlled randomised experiments. The adage 'all models are wrong; some models are useful' is applicable. Can the causal effect be estimated 'well enough', inclusive of both bias and efficiency, to improve understanding of the world, or to enable rational decisions to be made regarding policies, programmes or interventions? Perfection is out of reach, so utility must suffice.

Discussion

Figure 2.11 provides a summary of the results of applying the conditioning strategies discussed in the chapter to the simulated data set. Whereas the unconditional model is badly biased, markedly departing from the true value of the causal effect of A on C, all of the active conditioning methods have captured the true value within their 95% CIs.

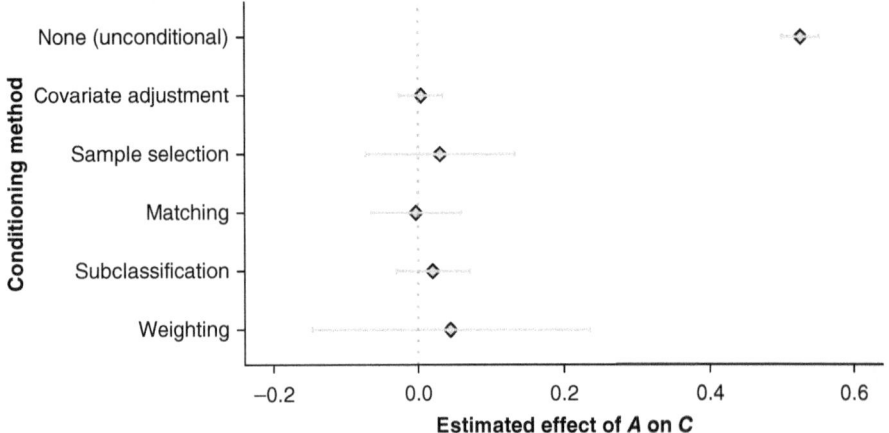

Figure 2.11 Point estimates and 95% confidence intervals by conditioning method. The dotted vertical line indicates the true value of the causal effect

As a conditioning strategy, covariate adjustment has some properties that set it apart from sampling, subclassification, matching or weighting. It is extremely simple to implement and, in general, produces the lowest variance (e.g. smallest standard error) estimates when compared to the other approaches. This is reflected in Figure 2.11 as the width of the 95% CIs.

The reason for this efficiency is that the parametric assumptions underlying the linear models that accomplish the covariate adjustment are a form of information that is brought to bear on the estimation process; information that does not come from the data (Berk, 2004). The problem, of course, is the potential that this information is incorrect. The analyst may inappropriately fit a model assuming that all of the effects are linear and non-interactive. Biased and inconsistent estimates would result.

The conditioning accomplished by covariate adjustment is done by projecting the statistical model's response surface across the multivariate distribution of the predictor variables. At times, this may involve extrapolating the solution across a region devoid of any data. In this case, the results are essentially pure speculation made possible by the model's strong assumptions. Figure 2.12 illustrates such a situation.

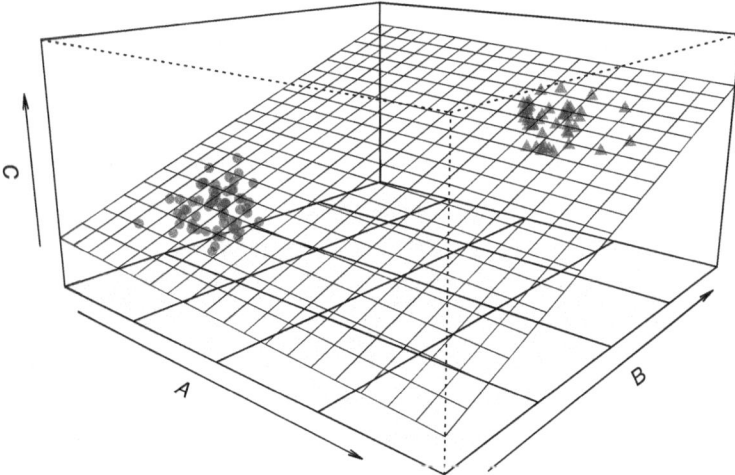

Figure 2.12 Conditioning via covariate adjustment can involve extensive extrapolation across regions of sparse or absent data

One can easily imagine absurd or nonsensical scenarios in which linear models will happily produce results – most likely, meaningless results. This can happen in situations where the 'treated' and 'control' subjects are extremely dissimilar with respect to the conditioning variables. The estimated effect of the treatment would be interpreted as the effect of treatment disallowing any variation on the conditioning variables, but the data set might not contain any subjects for whom treatment status varies independently of the background variables.

This situation can be relatively difficult to detect using covariate adjustment, but would be quite easy to discover using an approach like matching, subclassification or weighting. Using matching, one would notice that little to none of the sample would be matched within a reasonably sized caliper. With subclassification, many of the strata would be devoid of control (or treatment) subjects. And with weighting, the weights required to equalise the densities would be extreme.

The 'curse of dimensionality'

The general problem of algorithms becoming exponentially more intractable or computationally expensive as the number of variables or dimensions increases is known as the *curse of dimensionality* (Bellman, 1961). Analysts will typically need to condition on multiple variables in order to remove confounding bias. For example, Rosenbaum and Rubin (1983) conditioned on 74 variables to estimate the causal effect of coronary bypass surgery versus medication on post-heart attack survival.

Covariate adjustment is perhaps the only conditioning strategy considered here that is suitable for conditioning on multiple variables simultaneously.

Consider how difficult it would become to identify close matches on a large set of continuous variables. The difficulty scales geometrically with the number of variables (Ho et al., 2007). The curse of dimensionality affects the sample selection, weighting and subclassification strategies even more severely. Intentional conditioning via sample selection on a single variable already involves, in many cases, the discarding of vast amounts of data that would otherwise inform the estimation. The proportion of potential data that simultaneously satisfies multiple criteria is far smaller, and plummets geometrically as additional restrictions are introduced. A similar situation occurs with subclassification. Stratifying on multiple variables would involve populating the cells of an n-way contingency table (Kish & Anderson, 1978) Massive data availability in the form of sample sizes rarely observed in the social sciences would become necessary. Finally, consider the problem of weighting. Calculating weights requires density estimation, but multivariate density estimation is extraordinarily difficult when there are more than a few variables (Scott, 2015).

As a practical matter, when a researcher wishes to condition on multiple variables, the options are to either reluctantly accept the assumption-laden nature of covariate adjustment or break the curse of dimensionality by reducing the multiple variables down to one and then using matching, subclassification or weighting to do the conditioning. One such single-variable summary is called a **propensity score** (Guo & Fraser, 2015; Rosenbaum & Rubin, 1983). Propensity score methods of inference are discussed in Chapter 5.

In all cases, researchers must be ready and willing to abandon a project when circumstances indicate that inference is impossible or excessively leveraged on untenable assumptions. In those conditions, the data are simply unable to support the causal estimate that the researcher wishes to produce.

Further Reading

Berk, R. A. (2004). *Regression analysis: A constructive critique.* Sage.
This book introduces linear regression from a sceptical position, illustrating many common errors and misunderstandings and describing how they can be remedied. Berk is particularly critical of linear regression's application to causal inference, arguing that it is more likely to be useful as a descriptive tool for data reduction.

Fox, J. (2008). *Applied regression analysis and generalized linear models* (3rd ed.). Sage.
This is one of the most complete and thorough treatments of linear regression that has ever been written.

Freedman, D. A. (1991). Statistical models and shoe leather. *Sociological Methodology, 21*, 291–313.
This highly readable and entertaining paper contrasts causal inference via the thoughtless application of statistical models with a laborious, design-based approach, using Snow's work on identifying the cause of the 1853-54 London cholera epidemic as a motivating example.

Westfall, J., & Yarkoni, T. (2016). Statistically controlling for confounding constructs is harder than you think. *PLOS ONE 11*(3), e0152719. doi:10.1371/journal.pone.0152719
This paper illustrates the serious problems measurement error creates for causal inference via covariate adjustment.

3

DIRECTED ACYCLIC GRAPHS

Chapter Overview

A directed acyclic graph (DAG), also known as a causal diagram, is a representation of a researcher's theory regarding the causal relationships between variables involved in a particular area of research (Pearl et al., 2016). In general, statistical tools or techniques for doing causal inference can be divided into two types: those that require a fully elaborated DAG and those that require no DAG (or perhaps only a minimally elaborated DAG). Researchers in many disciplines of the social sciences may be unfamiliar with DAGs in spite of what I believe to be their central role in inference. This is a truly unfortunate state of affairs. Happily, interest and awareness of DAGs has been increasing in recent years, and I believe that concept will soon be viewed as an essential idea in applied statistics and research methodology.

Figure 3.1 is an example of a DAG. In DAGs, directed arrows (called 'edges') represent causal relationships between variables (called 'nodes'). In this DAG, there are three nodes: X, Y and Z. Here, Z is a cause of both X and Y, and X is also a cause of Y.

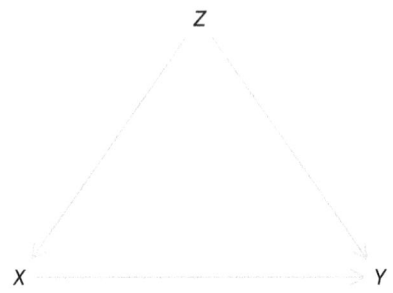

Figure 3.1   An example of a DAG

Note. DAG = directed acyclic graph.

One requirement of DAGs is that the edges connecting variables have definite direction, that is, they are *directed*. Another requirement is that there can be no feedback loops. Figure 3.2 displays a DAG with a feedback loop. This DAG cannot be analysed. Therefore, DAGs must be *acyclic*. Hence the term *directed acyclic graph*.

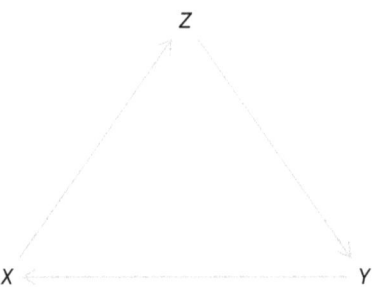

Figure 3.2   An improper DAG containing a feedback loop

There are cases when scientists might be interested in phenomena that exhibit feedback loops, which cannot be represented in a cross-sectional DAG. In many cases, these can be modelled by considering the temporal ordering of events explicitly. For example, X measured at occasion 1 could affect Y at occasion 2, which affects X again at occasion 3. This would be a permissible acyclic DAG because no observation of X ever causes itself temporaneously.

DAGs are similar to the path diagrams that one encounters in path analysis or structural equation modelling, but there is one *crucial* difference: in DAGs, no functional form is implied by an edge connecting variables, whereas in path diagrams, arrows represent regression coefficients describing linear structural relationships. An edge connecting nodes in a DAG simply means that the downstream variable relies, in some way, on the upstream variable for its value (Pearl et al., 2016).

DAGs are not path models

The causal relationships described by DAGs may be linear or non-linear, smooth or discontinuous. The model represented by a DAG communicates a smaller set of assumptions about the world than a similar path diagram or structural equation model. In those contexts, arrows represent coefficients in a particular statistical model, often with a linear functional form (Bollen, 2014). While DAGs are completely agnostic about the functional form of the causal relationships encoded by the arrows (edges), they do make definitive statements about statistical independence and dependence. They also describe exactly how dependence or independence between variables changes as a result of *conditioning* on variables in the graph (Greenland et al., 1999).

DAG terminology and variable roles

An ancestor is a node (or variable) that is located 'upstream' of a particular node. A descendant is located downstream (Pearl et al., 2016).

For example, the ancestors of node X in Figure 3.3 are E and B. Starting at a particular node of interest, ancestors include the set of nodes that can be reached by traversing the graph against the direction of the arrows. The descendants of X are A, C and Y; these are the nodes that can be reached by travelling with the direction of the arrows.

Because arrows represent causation in DAGs, a variable's ancestors are its causes. Its descendants are, in turn, the variables that it causes.

The ancestors of Y are C, B, X and E. However, node Y has no descendants. Nodes E, G, D and B have no ancestors; they are said to be **exogenous**. Any node with an

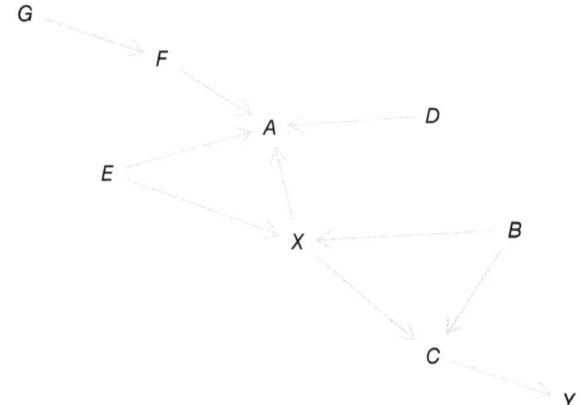

Figure 3.3 Example of a DAG illustrating ancestors and descendants

ancestor is **endogenous**. The endogenous nodes in Figure 3.3 are *F*, *A*, *X*, *C* and *Y*. A **chain** is a sequence of nodes connected by a set of unidirectional arrows. In other words, a chain is a set of nodes that can be traversed without the direction of the arrows ever changing. For example, $A \rightarrow B \rightarrow C \rightarrow D \rightarrow E$ is a chain of causation from *A* to *E*.

Figure 3.3 contains several chains, for example, $G \rightarrow F \rightarrow A$, $E \rightarrow X \rightarrow C \rightarrow Y$, $B \rightarrow X \rightarrow A$ and $B \rightarrow C \rightarrow Y$.

A **fork** occurs when a node has more than one directed edge emerging from it. In the DAG $A \leftarrow B \leftarrow C \rightarrow D \rightarrow E$, variable *C* forks into two chains, $C \rightarrow B \rightarrow A$ and $C \rightarrow D \rightarrow E$. Figure 3.3 contains forks originating at *E*, *B* and *X*.

A node's position in the graph with respect to ancestors, descendants and to the research question at hand determines how it must be treated. With practice, and for simple DAGs, these are easy to apprehend by visual examination. A later section introduces tools for automated DAG analysis.

Exposure

The exposure variable is the node whose causal effect is the focus of the research project (Shrier & Platt, 2008). For this reason, exposure variables are sometimes called focal variables or treatment variables. They are also known as independent variables or *X*-variables. This book will use all of these terms interchangeably. Some studies will have more than one exposure variable. For example, studies investigating mediation will include both the putative exogenous cause and one or more mediators as exposure variables. The identification of the exposure variable on a DAG is not based on its position with respect to ancestors and descendants, but rather on the hypothesis being investigated. In other words, locating the exposure variable(s) on the DAG requires knowledge from outside the DAG. When DAG nodes are identified with letters instead of variable names, X identifies the exposure variable by convention.

There are times when researchers may wish to make inferences about the effects of more than one variable in a DAG. In this case, there are multiple exposure variables, each of which will need to be considered separately. It is likely that estimating the causal effect of each of these exposure variables will require a different set of conditioning variables, implying that this generally cannot be accomplished with a single statistical analysis (Westreich & Greenland, 2013).

Outcome

The outcome variable is expected to change its value in response to change in the exposure variable. Following Pearl et al.'s (2016) brief definition of causality, the researcher hypothesises that the outcome variable may depend on the exposure for its value. The purpose of the study is to estimate the existence and magnitude of the causal effect of the exposure on the outcome. Outcome variables are also known as response variables, dependent variables or Y-variables. Like exposure variables, the identity of the outcome depends on the research context. It is conventionally represented by the letter Y unless full variable names are given.

Mediator

A mediator has the exposure variable as an ancestor and the outcome variable as a descendant (VanderWeele, 2009). It is therefore located in the midst of a chain connecting these nodes. A mediator is an intervening variable through which some or all of the causal effect of the exposure variable on the outcome is transmitted.

In Figure 3.4, nodes A and B are mediators. (Also note that this DAG is a chain.) In Figure 3.3, node C is a mediator of the effect of X on Y.

$$X \dashrightarrow A \dashrightarrow B \dashrightarrow Y$$

Figure 3.4 Example of a DAG with mediation

Confounder

A **confounder** is a node at the origin of a fork that includes both the exposure and the outcome in separate chains (Greenland et al., 1999). Confounders are ancestors of both the exposure and the outcome variables, but the confounder's effect on the outcome variable cannot be completely mediated by the exposure variable. In other words, a confounder must have an effect of the outcome that is not transmitted through the exposure. So A is not a confounder in the DAG $A \rightarrow X \rightarrow Y$ because A has no effect on Y that does not pass through X, even though A is an ancestor of X and Y.

In Figure 3.1, X is the exposure and Y is the outcome, following the usual conventions. Node Z is a confounder. In Figure 3.3, B is a confounder. Even though it does not have a direct edge to the outcome variable Y, its influence passes through C to Y.

Confounders create statistical dependence between the exposure and outcome variables that is unrelated to any causal link between them. When this dependence is linear and both the exposure and the outcome variables are continuous, this dependence will manifest as a disturbed correlation between the variables. Confounders are the primary reason for the adage 'correlation does not imply causation'.

In Figure 3.1, some portion of the statistical dependence of Y on X is because X does, in fact, cause Y. This is manifest as the edge connecting X to Y. However, this dependence will be altered by the confounding effect of variable Z. The magnitude or strength of the causal connection between X and Y cannot be correctly estimated from the data without conditioning on the confounder in some way. The same is true of Figure 3.3. The existence and strength of the causal relationship between X and Y could only be estimated in an unbiased manner after conditioning on B. More details on this point will come later in this chapter.

Proxy confounder

A proxy confounder is a variable within a chain leading from a confounder to either the exposure or the response nodes. It is a mediator of confounding – a variable through which the causal effect of a confounder passes on its way to the exposure or outcome variable (VanderWeele, 2019).

In Figure 3.5, Z and B are confounders because they originate at forks that include both X and Y. Node A is a proxy confounder, as it is a mediator of Z's effect on X. Node B, in addition to being a confounder in its own right, is also a proxy confounder of Z.

Instrument

An instrument, also known as an instrumental variable, is an ancestor of the exposure that is not a confounder or proxy confounder (Hernán & Robins, 2006). This means that the instrument can only be connected to the response through the exposure. In other words, the exposure completely mediates the effect of the instrument on the response. If the instrument had any connection to the response node that did not pass through the exposure, it would be a confounder instead of an instrument.

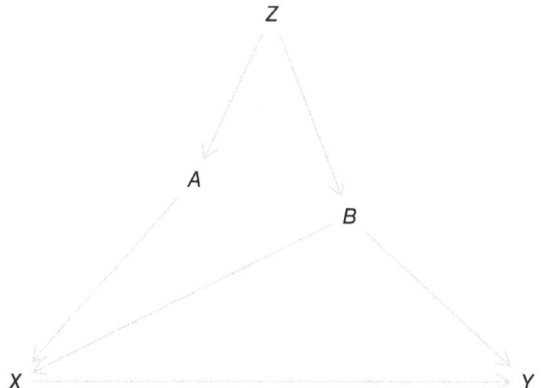

Figure 3.5 Example of a DAG with proxy confounders. *Z* and *B* are confounders. *A* and *B* are proxy confounders of *Z*

In Figure 3.6, node *A* is an instrument. When an instrument is available, researchers can use instrumental variable methods to estimate the causal effect without conditioning on any of the confounders. In fact, researchers do not even need to identify or measure confounding variables when an instrument is available. Unfortunately, instruments are generally hard to come by in the social sciences, but when available, they are extraordinarily valuable tools for estimating causal effects. Instrumental variables estimation is discussed in detail in Chapter 6.

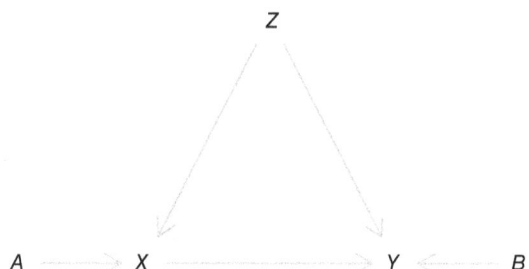

Figure 3.6 Example of a DAG with an instrument

Competing exposure

A competing exposure is an ancestor of the outcome variable but not the exposure (Law et al., 2012). In Figure 3.6, *B* is a competing exposure. Conditioning on competing exposures can increase the precision of estimation of causal effects, and also increases the statistical power of corresponding hypothesis tests. However, it is not necessary to conditioning on competing exposure variables to obtain unbiased causal estimates.

Collider

A collider or collision node has both the exposure and the outcome variable as ancestors; in other words, it is 'downstream' of both (Shrier & Platt, 2008).

In Figure 3.7, *A* is a collider. (Node *B* is a competing exposure; node *Z* is a confounder.)

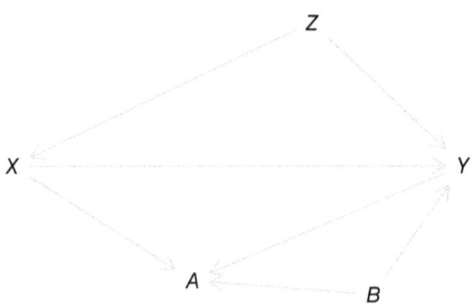

Figure 3.7   Example of a DAG with a collider

d-separation, *d*-connectedness and statistical independence

A pair of nodes are said to be *d*-connected when an unblocked connecting path exists between them. (The *d* means *directional*.) Colliders block *d*-connectedness, and so does conditioning on any intermediate nodes in the chain. Similarly, a pair of nodes is *d*-separated when no connecting path exists between them, or when every path connecting them is *blocked* (Pearl, 1995).

When nodes are *d*-connected, they will be statistically dependent. When they are *d*-separated, they will be statistically independent. Statistical independence implies no correlation between variables, in fact, no relationship of any kind (DeGroot & Schervish, 2002).

Figure 3.8 provides another example of a DAG containing a collider.

Figure 3.8   Another example of a DAG containing a collider

Node *C* is a collider. The node pairs *AB*, *AC*, *BC*, *CD* and *DE* are *d*-connected and therefore will be statistically dependent. Node pairs *AD*, *AE*, *BD* and *BE* are *d*-separated and statistically independent due to the collider. Note that the observed correlation between the *d*-connected nodes can be any value, even zero, in spite of the statistical dependence because their relationship need not be linear (and correlation coefficients detect linear associations). However, the correlation between the *d*-separated nodes *must* be zero.

This pattern of statistical dependence and independence implied by the DAG can be verified through simulation. I simulated a data set from the DAG in Figure 3.8 and calculated a correlation matrix, which is displayed in Table 3.1. All the relationships in the simulated data are linear, so Pearson correlation coefficients provide a good summary of the relationship between variables.

Table 3.1 Correlation matrix for data simulated from Figure 3.7

	A	B	C	D
A	1.000			
B	.708	1.000		
C	.451	.634	1.000	
D	.006	−.004	.622	1.000
E	.005	.004	.434	.699

The correlations between the d-connected nodes are as follows: r_{AB} = .708, r_{AC} = .451, r_{BC} = .634, r_{CD} = .622, r_{CE} = .434 and r_{DE} = .699. All of these have sizable positive correlations as a consequence of their d-connectedness.

In contrast, the correlations between the d-separated nodes from the DAG in Figure 3.8 are as follows: r_{AD} = .006, r_{AE} = .005, r_{BD} = −.004 and r_{BE} = .004. These are essentially zero, and would be exactly zero if not for sampling error.

Conditioning

Conditioning also blocks d-connectedness. In Figure 3.8, the chain $A \rightarrow B \rightarrow C$ is d-connected (as is chain $C \leftarrow D \leftarrow E$). Conditioning on B will result in A and C becoming d-separated and therefore independent. When independence is created via conditioning, the term **conditionally independent** is used. The node A is d-separated and conditionally independent of C after conditioning on B.

As described in Chapter 2, there are many ways of conditioning. Given that covariate adjustment via a linear regression model (a) can be accomplished in a single step, (b) is readily applicable when there are multiple variables to condition on, (c) works well when its assumptions are met and (d) is likely to be familiar to most readers, I will adopt it as the default method throughout this chapter. First, I will fit a model regressing A on C with no conditioning using the lm() function in R. The model is $A_i = b_0 + b_1(C_i) + e_i$. Here is the output:

```
##
## Call:
## lm(formula = A ~ C)
```

```
##
## Residuals:
##      Min      1Q  Median      3Q     Max
## -3.7506 -0.6103 -0.0067 0.6094 3.2334
##
## Coefficients:
##                Estimate Std. Error t value Pr(>|t|)
## (Intercept) -0.004044   0.009037  -0.447    0.655
## C            0.204310   0.004045  50.504   <2e-16 ***
## ---
## Signif. codes: 0 '***' 0.001 '**' 0.01 '*' 0.05 '.' 0.1 ' ' 1
##
## Residual standard error: 0.9037 on 9998 degrees of freedom
## Multiple R-squared:  0.2033, Adjusted R-squared:  0.2032
## F-statistic:  2551 on 1 and 9998 DF,  p-value: < 2.2e-16
```

It is clear that A is dependent on C, $b_1 = 0.204$ (0.004), 95% CI [0.196, 0.212], $p < .001$.

Now I will condition on B by adding it to the regression model as an additional predictor. The new model is $A_i = b_0 + b_1(C_i) + b_2(B_i) + e_i$. The effect of C will be estimated after controlling for B. (The effect of B will also be estimated after statistically controlling for C, but that is not the focus.)

```
##
## Call:
## lm(formula = A ~ C + B)
##
## Residuals:
##      Min       1Q   Median       3Q      Max
## -2.64556 -0.48039 -0.00021  0.47727  2.43153
##
## Coefficients:
##                Estimate Std. Error t value Pr(>|t|)
## (Intercept) -0.007032   0.007145  -0.984    0.325
## C            0.001211   0.004137   0.293    0.770
## B            0.501105   0.006472  77.429   <2e-16 ***
## ---
## Signif. codes:  0 '***' 0.001 '**' 0.01 '*' 0.05 '.' 0.1 ' ' 1
##
## Residual standard error: 0.7145 on 9997 degrees of freedom
## Multiple R-squared:  0.5019, Adjusted R-squared:  0.5018
## F-statistic:  5038 on 2 and 9997 DF,  p-value: < 2.2e-16
```

The results indicate that A is conditionally independent of C after conditioning on B, $b_1 = 0.001$ (0.004), 95% CI [–0.007, 0.009], $p = .77$. This is because B completely mediates the causal effect of A on C. After conditioning on B, this path is blocked, and A and C become d-separated.

From the same output, it is clear that A is still dependent on B after conditioning on C. This is understandable from the DAG, as the causal effect of A on B is not transmitted through C. Conditioning on C does not affect the statistical relationship between A and B.

Conditioning on colliders

As shown in Figure 3.8, the nodes on the left side of C are statistically independent from the nodes on the right side of C, because colliders block d-connectedness. But conditioning on the collider or any of its downstream descendants removes the blockage (Rohrer, 2018). We have already seen that the node pair AE are independent according to the near-zero correlation coefficient.

First, A is regressed on E, demonstrating that they are independent prior to conditioning on C. The regression model is $A_i = b_0 + b_1(E_i) + e_i$.

```
##
## Call:
## lm(formula = A ~ E)
##
## Residuals:
##      Min      1Q   Median      3Q     Max
## -3.6702 -0.6643 -0.0100 0.6846 3.8105
##
## Coefficients:
##                Estimate Std. Error t value Pr(>|t|)
## (Intercept) -0.006516    0.010124  -0.644    0.520
## E            0.004951    0.010219   0.484    0.628
##
## Residual standard error: 1.012 on 9998 degrees of freedom
## Multiple R-squared: 2.347e-05, Adjusted R-squared:  -7.654e-05
## F-statistic: 0.2347 on 1 and 9998 DF,  p-value: 0.6281
```

The results indicate a near-zero and non–statistically significant relationship between A and E before conditioning, $b_1 = 0.005$ (0.01), 95% CI [–0.015, 0.025], $p = .628$.

Next, a second regression model is fit to the data, this time adding the collider C as a conditioning variable. The new model is $A_i = b_0 + b_1(E_i) + b_2(C_i) + e_i$.

```
##
## Call:
## lm(formula = A ~ E + C)
```

```
##
## Residuals:
##     Min       1Q  Median       3Q      Max
## -3.7461 -0.6021 -0.0031  0.5860  3.1614
##
## Coefficients:
##               Estimate Std. Error t value Pr(>|t|)
## (Intercept) -0.004486   0.008779  -0.511    0.609
## E           -0.240366   0.009836 -24.436   <2e-16 ***
## C            0.250595   0.004363  57.441   <2e-16 ***
## ---
## Signif. codes:  0 '***' 0.001 '**' 0.01 '*' 0.05 '.' 0.1 ' ' 1
##
## Residual standard error: 0.8779 on 9997 degrees of freedom
## Multiple R-squared:  0.2482, Adjusted R-squared:  0.248
## F-statistic:  1650 on 2 and 9997 DF,  p-value: < 2.2e-16
```

Now A and E are statistically dependent, $b_1 = -0.24$ (0.01), 95% CI [–0.26, –0.221], $p < .001$. If a researcher was interested in studying the relationship between these two variables, and either accidentally or intentionally conditioned on C, their conclusion would be wrong.

Colliders and the real world

Colliders exhibit themselves frequently in real-world contexts, and researchers rarely seem to notice them. They become a particular problem when researchers unknowingly condition on them. For example, a great deal of psychology studies have been based on college student samples. Many colleges have selective admissions criteria. Students must meet minimal academic qualifications in order to be admitted. However, colleges also recruit outstanding athletes with lower academic qualifications. The situation could be represented by the DAG in Figure 3.9.

Athletic_achievement   Academic_achievement

College_admission

Figure 3.9   DAG for selection on college admission example

Sampling – or to be more specific, non-probability sampling – is a conditioning method. By using college student samples, psychologists have (perhaps unknowingly)

conditioned on college admission. Imagine that researchers seek to test the 'nerd hypothesis': that high academic achievement causes low athletic ability. They collect data from a large sample of college students. Further, assume for the sake of argument that the DAG in Figure 3.9 is correct. According to this DAG, athletic achievement does not cause academic achievement. No arrow connects them. They are d-separated and therefore independent.

However, by collecting data from college students, the researchers have inadvertently conditioned on college admission, a collider. This creates an association between athletic achievement and academic achievement that does not otherwise exist. If the probability of college admission is positively related to both athletic and academic achievement, the observed correlation between athletic and academic achievement will become negative. It is likely that an incorrect, misleading conclusion will be reached: that academic achievement causes a reduction in athletic ability. Even if the researchers do not interpret the results as causal, they may still conclude that athletics and academics are negatively correlated. The confirmation of the 'nerd hypothesis' would undoubtedly receive breathless coverage in the popular press.

This can be readily observed in simulation. I simulated another data set in accordance with the DAG in Figure 3.9. The data set contains an indicator variable for college admission.

First, the complete data set (including college and non-college students) is analysed via a simple linear regression of *academics* on *athletics* via a regression model $athletic_i = b_0 + b_1 (academic) + e_i$. The output is as follows:

```
##
## Call:
## lm(formula = athletic ~ academic, data = data)
##
## Residuals:
##     Min      1Q  Median      3Q     Max
## -2.9883 -0.6851 -0.0219  0.6971  3.8133
##
## Coefficients:
##              Estimate Std. Error t value Pr(>|t|)
## (Intercept) -0.01154    0.03275  -0.353    0.725
## academic     0.00637    0.03150   0.202    0.840
##
## Residual standard error: 1.035 on 998 degrees of freedom
## Multiple R-squared: 4.098e-05, Adjusted R-squared:  -0.000961
## F-statistic: 0.0409 on 1 and 998 DF, p-value: 0.8398
```

The analysis indicates no evidence of any association between athletics and academics, b_1 = 0.006 (0.031), 95% CI [–0.055, 0.068], p = .84. This model is correctly estimating the true, nil causal effect. It has reached the correct conclusion.

Next, the subset of college students is analysed via the same model. This analysis is implicitly conditioned on college attendance, the collision node, via the sampling process.

```
##
## Call:
## lm(formula = athletic ~ academic, data = data.college)
##
## Residuals:
##     Min      1Q  Median      3Q     Max
## -2.6726 -0.6348 -0.0423  0.5010  3.5922
##
## Coefficients:
##              Estimate Std. Error t value Pr(>|t|)
## (Intercept)   0.57098    0.04339   13.16  < 2e-16 ***
## academic     -0.26324    0.04259   -6.18  1.3e-09 ***
## ---
## Signif. codes:  0 '***' 0.001 '**' 0.01 '*' 0.05 '.' 0.1 ' ' 1
##
## Residual standard error: 0.8784 on 517 degrees of freedom
## Multiple R-squared: 0.0688, Adjusted R-squared: 0.067
## F-statistic: 38.2 on 1 and 517 DF, p-value: 1.298e-09
```

This analysis finds a negative association between athletics and academics, $b_1 = -0.263(0.043)$, 95% CI [-0.347, -0.18], $p < .001$. However, in substantive terms, this result is completely misleading. Academic ability and athletic ability appear to be negatively related when, in reality, they are not causally linked – a consequence of conditioning on the collider. Figure 3.10 displays the scatter plots and regression lines corresponding with the regression results.

Spurious paths

A spurious path is a connection between the exposure and outcome variables originating outside of the chain linking the exposure to the response, if one exists (Hernán et al., 2004). Spurious paths are created by confounders. When unblocked spurious paths exist, the unadjusted statistical relationship between the exposure and outcome do not measure the strength of any existing causal link between them.

Figure 3.11 provides an example of such a DAG. Here, Z is a confounder and A and B are proxy confounders. The exposure X and outcome Y are d-connected, because they are connected by an unblocked path originating at the confounder Z. Remember, colliders and conditioning are the two conditions that create blocking.

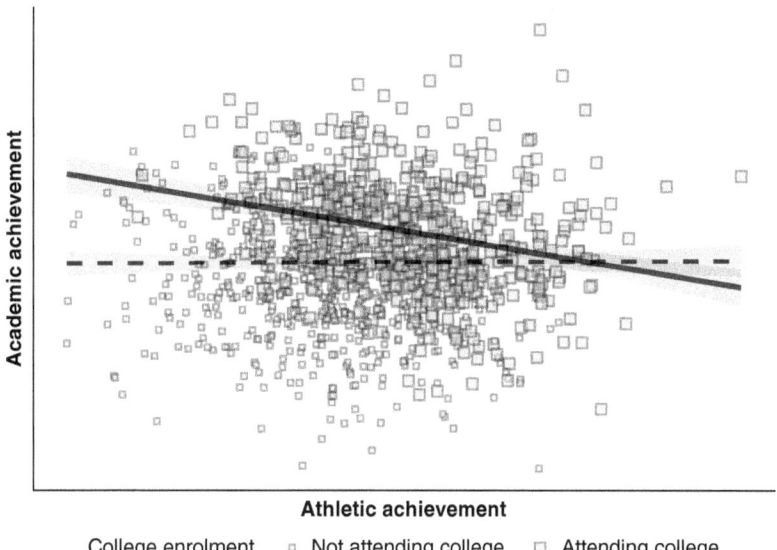

Figure 3.10 Bias from conditioning on a collider. The regression line fitted to only the college attendees (solid line) has a negative slope. The regression line fitted to all of the data (dashed line) has the correct zero slope

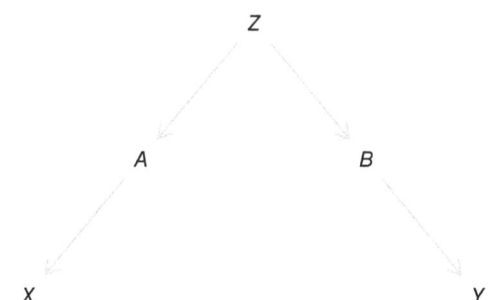

Figure 3.11 DAG containing a spurious path

In this DAG, X does not cause Y. However, X and Y are statistically dependent due to their d-connectedness. It is very likely that X and Y will be correlated as a result. Situations like the one shown in Figure 3.11 exemplify the meaning of the phrase 'correlation does not imply causation.' The statistical dependence resulting from the spurious path can be demonstrated via analysis of simulated data generated from the DAG. A simple linear regression model of form $Y_i = b_0 + b_1(X_i) + e_i$ is fit to the simulated data. This model does not condition on any of the other variables. As a result, the spurious path connecting X and Y is unblocked.

```
##
## Call:
## lm(formula = Y ~ X)
##
## Residuals:
##     Min      1Q Median     3Q     Max
## -6.6240 -1.0946 0.0085 1.0804 5.2052
##
## Coefficients:
##               Estimate Std.  Error t value Pr(>|t|)
## (Intercept) -0.04413    0.04161  -1.061   0.289
## X            0.31917    0.02261  14.118   <2e-16 ***
## ---
## Signif. codes:  0 '***' 0.001 '**' 0.01 '*' 0.05 '.' 0.1 ' ' 1
##
## Residual standard error: 1.611 on 1498 degrees of freedom
## Multiple R-squared: 0.1174, Adjusted R-squared:  0.1168
## F-statistic: 199.3 on 1 and 1498 DF, p-value: < 2.2e-16
```

The results reveal the statistical dependence, $b_1 = 0.319$ (0.023), 95% CI [0.275, 0.363], $p < .001$. An unsophisticated analyst or consumer of research might interpret this result as causal, or potentially causal. But this is clearly not the case.

Spurious paths, like all paths, can be blocked by conditioning. Based on this DAG, the analyst could choose to condition on A, B or Z. It is instructive to consider how the DAG from Figure 3.11 would change in response to each conditioning decision. Figure 3.12 presents these modified DAGs.

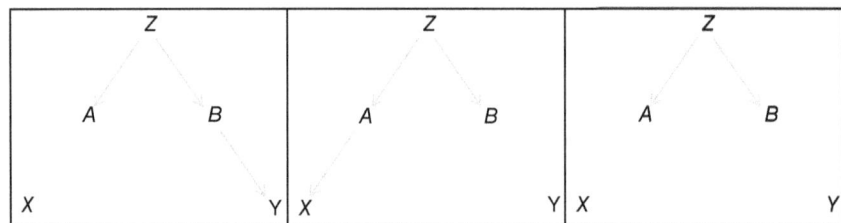

Figure 3.12   Effect of conditioning. Left panel: Conditioning on A. Centre panel: Conditioning on B. Right panel: Conditioning on Z. Any of these conditioning decisions block the spurious connection between X and Y and would permit the unbiased estimation of their non-existent causal relationship

Node A is conditioned on in the left panel, blocking the $Z \rightarrow X$ path that passes through it. This path has been removed from the DAG. Conditioning on A blocks the confounding path from Z to X. After conditioning, Z and B become competing exposures (with B mediating the effect of Z); X and Y become

d-separated. Fitting a model conditioning on A, $Y_i = b_0 + b_1(X_i) + b_2(A_i) + e_i$ demonstrates that once A is conditioned on, the estimated relationship between X and Y approaches zero and becomes non-significant.

```
## 
## Call:
## lm(formula = Y ~ X + A)
## 
## Residuals:
##     Min      1Q  Median      3Q     Max
## -6.0113 -1.0838 -0.0035  1.0686  4.7097
## 
## Coefficients:
##               Estimate Std. Error t value Pr(>|t|)
## (Intercept) -0.04020    0.04053  -0.992    0.322
## X            0.01970    0.03979   0.495    0.621
## A            0.44565    0.04932   9.037   <2e-16 ***
## ---
## Signif. codes:  0 '***' 0.001 '**' 0.01 '*' 0.05 '.' 0.1 ' ' 1
## 
## Residual standard error: 1.57 on 1497 degrees of freedom
## Multiple R-squared:  0.1631, Adjusted R-squared:  0.162
## F-statistic: 145.9 on 2 and 1497 DF,  p-value: < 2.2e-16
```

Node B has been conditioned on in the centre panel of Figure 3.12. This blocks the $Z \rightarrow B \rightarrow Y$ confounding path, converting Z and A to instruments. Linear regression results demonstrate, once again, that the estimated relationship between X and Y is essentially zero after conditioning on B.

```
## 
## Call:
## lm(formula = Y ~ X + B)
## 
## Residuals:
##     Min      1Q  Median      3Q     Max
## -3.1519 -0.6882  0.0240  0.6937  3.6680
## 
## Coefficients:
##                Estimate Std. Error t value Pr(>|t|)
## (Intercept) -0.0168122  0.0264461  -0.636    0.525
## X            0.0003949  0.0158838   0.025    0.980
## B            0.9744187  0.0207145  47.040   <2e-16 ***
## ---
## Signif. codes: 0 '***' 0.001 '**' 0.01 '*' 0.05 '.' 0.1 ' ' 1
```

```
##
## Residual standard error: 1.024 on 1497 degrees of freedom
## Multiple R-squared: 0.6439, Adjusted R-squared:  0.6434
## F-statistic: 1353 on 2 and 1497 DF, p-value:  < 2.2e-16
```

Node Z has been conditioned on in the right panel of Figure 3.12. This breaks the spurious path, converting A to an instrument and B to a competing exposure. As shown below, linear regression results indicate that conditioning on Z is also effective.

```
##
## Call:
## lm(formula = Y ~ X + Z)
##
## Residuals:
##      Min      1Q  Median      3Q     Max
## -5.3530 -0.9255 -0.0262  0.9520  4.8690
##
## Coefficients:
##               Estimate Std. Error t value Pr(>|t|)
## (Intercept) -0.035324   0.035979  -0.982    0.326
## X           -0.006084   0.024309  -0.250    0.802
## Z            0.981615   0.043611  22.508   <2e-16 ***
## ---
## Signif. codes:  0 '***' 0.001 '**' 0.01 '*' 0.05 '.' 0.1 ' ' 1
##
## Residual standard error: 1.393 on 1497 degrees of freedom
## Multiple R-squared:  0.3406, Adjusted R-squared:  0.3397
## F-statistic: 386.6 on 2 and 1497 DF,  p-value: < 2.2e-16
```

Researchers might choose to condition on combinations of these variables: for example, AB, AZ, BZ or ABZ. Any set of them will work.

As shown in Table 3.2, the estimated relationship of X and Y is essentially zero regardless of which set of variables is conditioned on, so long as this blocks the spurious path between X and Y. There is no problem of 'overcontrol' from conditioning on too many variables (Elwert & Winship, 2014).

Table 3.2 Linear regression model results: Estimated effect of X on Y after conditioning on confounders or proxy confounders

Conditioning set	Estimate	SE	t-Value	p-Value
(None)	0.319	0.023	14.118	<.001
A	0.020	0.040	0.495	.621
B	0.000	0.016	0.025	.980
Z	−0.006	0.024	−0.250	.802

Conditioning set	Estimate	SE	t-Value	p-Value
AB	0.019	0.026	0.745	.456
AZ	0.016	0.035	0.463	.643
BZ	−0.007	0.018	−0.378	.705
ABZ	0.019	0.026	0.738	.461

Unobservables

An **unobservable** is simply a variable that has not been measured and is therefore unavailable for inclusion in the set of conditioning variables (Greenland et al., 1999). A DAG must include all of the variables thought to be involved in the causal system of the exposure's effect on the outcome, not only those that have been measured or could be measured (Rohrer, 2018). Unobservables are often represented by the letter U.

Unobservables can threaten the ability of researchers to make causal inferences. The DAG shown in Figure 3.13 contains an unobservable confounder U. The causal effect of X on Y cannot be identified through conditioning. The variable U must be conditioned on in order to interpret the relationship between X and Y as a valid measure of the causal effect.

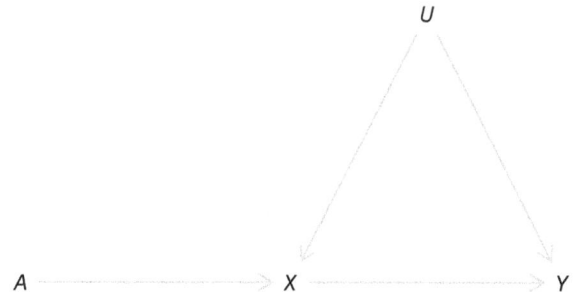

Figure 3.13 A DAG containing an unobservable (U) and an instrument (A). The causal effect of X on Y cannot be identified via conditioning

There is an alternative strategy that can be employed in the situation depicted by this DAG called instrumental variables analysis, which is the topic of Chapter 6 in this book.

Conditioning on mediators

Applied researchers are sometimes reluctant to condition on large sets of variables due to a fear of 'overcontrol'. Conditioning away effects of interest is not a problem of the size of the adjustment set, but rather its composition (Elwert & Winship, 2014). In other words, 'overcontrol bias' as understood by many applied researchers is a consequence of including the *wrong* variables in the adjustment set, not too many of the *right* ones.

In the DAG $X \rightarrow A \rightarrow Y$, node A completely mediates the causal effect of X on Y. Conditioning on A would render X and Y conditionally independent and therefore entirely remove the causal effect. Conditioning on mediators is the origin of the putative problem of overcontrol. In fact, researchers should generally avoid conditioning on any downstream descendant of the exposure variable. These must be either mediators or colliders, and conditioning on either biases estimated causal effects (Rohrer, 2018).

Consider the DAG in Figure 3.14. Nodes A and B are mediators of the causal effect of X on Y. There are no spurious paths.

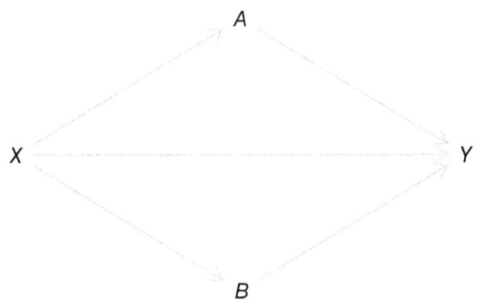

Figure 3.14　DAG containing direct and indirect effects

There are no spurious paths because there is no confounding; X and Y have no common ancestors. The nodes A and B are mediators of the causal effect of X on Y. The causal effect of X on Y can be decomposed into three components:

1　The indirect effect mediated by A: $X \rightarrow A \rightarrow Y$
2　The indirect effect mediated by B: $X \rightarrow B \rightarrow Y$
3　The direct (unmediated) effect: $X \rightarrow Y$

The total effect of X on Y is the sum of the direct and indirect effects. In this DAG, no conditioning is required to estimate the total effect. To illustrate how various conditioning decisions affect the estimate, I simulated a data set from this DAG. The mediated effect is the product of the slope of the mediator's relationship with X and the slope of Y's relationship with the mediator (MacKinnon, 2012). In the simulation, both of these slope coefficients are set to 1.0. The direct effect of $X \rightarrow Y$ is also set to 1.0. Therefore, the total effect should be $\underset{\text{indirect via A}}{(1 \times 1)} + \underset{\text{indirect via B}}{(1 \times 1)} + \underset{\text{direct}}{(1)} = 3$. Fitting a simple regression model to the simulated data confirms that the estimated total effect of X on Y is, indeed, very close to 3.0, as shown in the following results.

```
##
## Call:
## lm(formula = Y ~ X)
##
## Residuals:
##     Min     1Q Median     3Q    Max
## -7.1732 -1.2126 0.0183 1.3376 5.7725
##
## Coefficients:
##             Estimate Std. Error t value Pr(>|t|)
## (Intercept)  0.01669    0.05767   0.289    0.772
## X            3.07918    0.05574  55.237   <2e-16 ***
## ---
## Signif. codes:  0 '***' 0.001 '**' 0.01 '*' 0.05 '.' 0.1 ' ' 1
##
## Residual standard error: 1.823 on 998 degrees of freedom
## Multiple R-squared: 0.7535, Adjusted R-squared: 0.7533
## F-statistic: 3051 on 1 and 998 DF, p-value: < 2.2e-16
```

This unconditional model estimates the **total effect** of X on Y, as it includes both the direct and the indirect effects. The direct effect is that portion of the causal effect of X on Y that is not mediated. An indirect effect is the portion that is mediated through a specific pathway, and the total effect is the sum of the direct and indirect effects (MacKinnon, 2012). The estimate produced by the model estimates how Y would change given a 1-unit manipulation of X. When conditioning on a mediator, the path from X to Y passing through that mediator is blocked. The estimated effect of X on Y in such a model is no longer the total effect, but becomes the effect of a 1-unit change of X on Y if one could prevent the conditioned-on mediator from changing.

Figure 3.15 displays the effects that are estimated when conditioning on variables A, B or AB. The blocked paths have been removed.

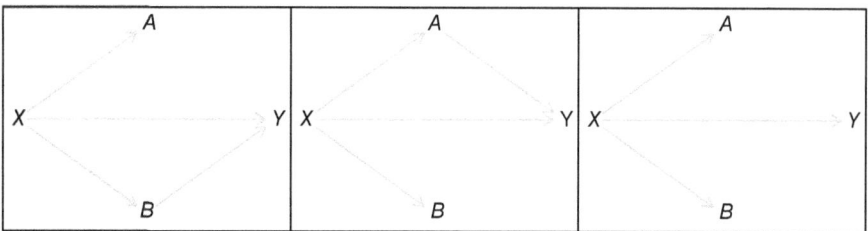

Figure 3.15 Effect of conditioning on mediators. Left panel: Conditioning on A. Centre panel: Conditioning on B. Right panel: Conditioning on A and B

A **direct effect** is the causal effect of X on Y that is transmitted directly, without passing through any mediators. In order to estimate the direct effect of X on Y given the DAG in Figure 3.14, it is necessary to condition on a set of variables such that all the mediating paths are blocked. Table 3.3 provides regression results from four linear regression models, each employing a different set of conditioning variables.

Table 3.3 Linear regression model results: estimated effect of X on Y after conditioning on mediators

Conditioning set	Estimated effect	Estimate	SE	t-Value	p-Value
(None)	Total	3.079	0.056	55.237	<.001
A	Direct + Indirect B	2.035	0.064	31.989	<.001
B	Direct + Indirect A	1.996	0.066	30.323	<.001
AB	Direct only	0.999	0.056	17.933	<.001

In most research contexts, the estimation of the total effect is the primary aim. Statistical mediation analysis is an exception, in which researchers attempt to decompose total effects into direct and mediated effects to understand causal mechanisms. Here, researchers intentionally attempt to isolate these effects using various conditioning decisions (MacKinnon, 2012). However, obtaining unbiased results from mediation analysis is generally much more difficult than estimating causal (total) effects of individual variables, because the answer now depends on estimating multiple coefficients correctly. Mediation analysis therefore has a high chance of producing erroneous or misleading results and should be used and interpreted with caution (Bullock et al., 2010).

Criteria for valid causal inference

There are two criteria described in Pearl's (1995) work on DAGs which, when satisfied, will permit causal inference in the absence of experimental manipulation: (1) the back-door criterion and (2) the front-door criterion.

Back-door criterion

The **back-door criterion** is a formalisation of the discussion of how causal inference requires all spurious paths to be blocked. Pearl et al. (2016) gave the following definition of the back-door criterion:

> Given an ordered pair of variables (X, Y) in a DAG, a set of variables Z satisfies the back-door criterion relative to (X, Y) if no node in Z is a descendant of X, and Z blocks every path between X and Y that contains an arrow into Y. (p. 61).

They further state as follows:

> In general, we would like to condition on a set of nodes Z such that
>
> 1   We block all spurious paths between X to Y.
> 2   We leave all directed paths from X to Y unperturbed.
> 3   We create no new spurious paths. (p. 61)

This is called the back-door criterion because of the paths that confounders have to the X variable. Pearl (2009) observed that 'these paths can be viewed as entering X_i through the back door' (p. 124).

Front-door criterion

An alternative criterion that makes possible the identification of causal effects is given by the front-door criterion. The DAG in Figure 3.16 displays what initially appears to be a hopeless situation, in which the causal effect $X \rightarrow Y$ is disturbed by unobservable confounder U. It is impossible to satisfy the back-door criterion because doing so would require conditioning on U, which is unavailable.

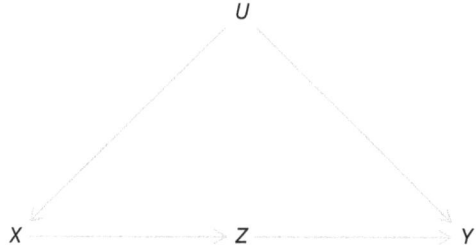

Figure 3.16   An illustration of the front-door criterion. U is an unobservable confounder

However, the causal effect of X on Y in this situation is identifiable via the front-door criterion, which Pearl et al. (2016) defined as follows:

> A set of variables Z is said to satisfy the front-door criterion relative to an ordered pair of variables (X, Y) if
>
> 1 Z intercepts all directed paths from X to Y.
> 2   There is no unblocked path from X to Z.
> 3   All back-door paths from Z to Y are blocked by X. (p. 69)

Criteria (1) and (2) imply that Z completely mediates the effect of X on Y. If this is the case, the causal effect can be calculated as a composite of the $X \rightarrow Z$ and $Z \rightarrow Y$ effects. The $X \rightarrow Z$ effect can be identified without conditioning because U is not a confounder of this relationship. The $Z \rightarrow Y$ path is confounded by U, but it can be identified by conditioning on X because X is a proxy confounder of U, and criterion (3) prohibits any other spurious paths from Z to Y. The product of these two values estimates the effect of X on Y (MacKinnon, 2012).

Situations in which the front-door criterion can be applied are rare. Pearl et al. (2016) give an example in which X is smoking, Z is tar deposits in the lungs and Y is lung cancer. The applicability of the front-door criterion hinges on the degree to which tar deposits in the lung can be measured, the possibility that the smoking to tar deposits path is itself unconfounded and whether tar deposits are the *only* way for smoking to cause lung cancer. If, for example, smoking can also cause lung cancer via toxins absorbed into the bloodstream, the front-door criterion would not be satisfied. Nor would it be satisfied if some confounder variable (perhaps exercise or genetics factors) influences both smoking behaviour and the ability of the lungs to clear tar deposits. The front-door criterion can be satisfied by multiple Z variables, but they are jointly subject to these restrictions.

Minimal and sufficient adjustment sets

A *sufficient adjustment set* contains a set of conditioning variables that satisfy the back-door criterion, such that all back-door paths from the exposure to the response are blocked (Greenland et al., 1999), but does not contain any downstream descendants of the exposure – which would be either mediators or colliders. In most DAGs, there are several sufficient adjustment sets. As the name implies, any of them will do the job. The graph in Figure 3.12 is a good example: the sufficient adjustment sets include A, B, Z, AB, AZ BZ and ABZ. The results presented in Table 3.2 confirm that conditioning on any of these sets allows for the correct estimation of the causal effect.

Conditioning on a sufficient adjustment set makes the causal effect identifiable, meaning that it can be estimated from the data.

A minimal adjustment set is a sufficient adjustment set that contains the smallest number of variables necessary to identify the causal effect (Greenland et al., 1999). The minimal adjustment sets for Figure 3.12 are A, B or Z. Conditioning on any of the sufficient adjustment sets will remove confounding bias from the causal effect estimate, so why are minimal adjustment sets preferred? Data collection is expensive, time-consuming and potentially invasive; there is little justification for collecting more auxiliary information than is strictly needed to address the core research questions. Second, an unnecessarily large set of conditioning variables increases the analyst's exposure to risk involving mistaken variable roles due to an incorrect DAG.

When there are multiple sufficient adjustment sets, the quality of measurement can provide a strong justification for choosing among the available options. Conditioning on imperfectly measured variables results in only partial blockage of confounding paths and incomplete fulfilment of the back-door criterion (Westfall & Yarkoni, 2016). Substantial bias in estimated effects can remain even after conditioning on a sufficient set of poorly measured variables.

Simultaneous estimation of causal effects

It is common for applied researchers to interpret multiple coefficients from a single regression (or related) model's results, as though each were unbiased estimates of causal effects. Westreich and Greenland (2013) referred to this as a 'Table 2 fallacy' since regression tables are commonly presented in Table 2 in epidemiology papers.

In general, a regression model can correctly estimate one and only one causal effect. Any additional variables included in the model serve the auxiliary purpose of statistical conditioning. Their only job is to assist in the unbiased estimation of the exposure variable's effect. As such, their coefficients have no causal interpretation, and therefore no meaning. Westerich and Greenland (2013) argue that the regression coefficients for conditioning variables should not even be presented in results tables, as this invites misinterpretation. I agree with this advice.

Figure 3.17 shows four potential situations. It is instructive to consider how the results of a multiple regression model, $Y_i = b_0 + b_1(X_i) + b_2(A_i) + e_i$, would be interpreted under each scenario.

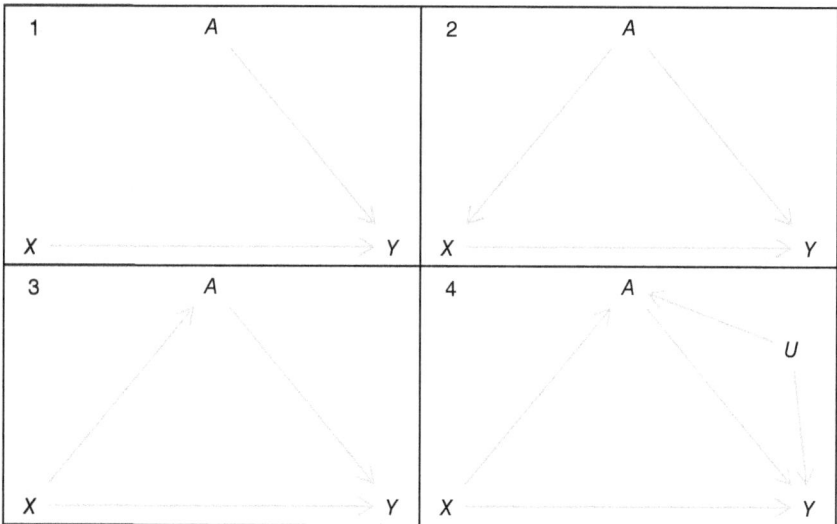

Figure 3.17 Simultaneous estimation of causal effects and potential DAGs. Panel 1: A and X as competing exposures. This is the only condition in which simultaneous estimation should be used. Panel 2: A as confounder of X → Y, X as mediator of A → Y. Panel 3: A as mediator of X → Y, X as confounder of A → Y. Panel 4: A as mediator of X → Y. X and U as confounders of A → Y. The causal effect A → Y is not estimable after conditioning on X due to U

In panel 1, A and X are independent, unconfounded competing exposures. This is the only situation in which X and A's causal effects can be estimated by a single model. Parameters b_1 and b_2 would correctly estimate their corresponding causal effects. Suffice it to say that I believe this situation to be very rare in the social sciences. It would occur if A and X were randomly assigned, but it is difficult for me to imagine many other plausible scenarios when this would occur.

In panel 2, A is a confounder of $X \rightarrow Y$, but X is a mediator of $A \rightarrow Y$. The regression model including A would correctly estimate the causal effect of X. But it would estimate only the direct effect, rather than the total effect, of A because X is in the model. Thus, the model results would consist of a total effect (for X, b_1) and a direct effect (for A, b_2). These two estimates cannot be interpreted on identical terms, they have fundamentally different meanings.

In panel 3, the situation is same as panel 2 with X and A's roles reversed. The model would estimate the total effect of A, (b_2) and the direct effect of X (b_1). These results would be reported in the same table of regression results and could be quite easily misinterpreted.

In panel 4, an unmeasured confounder U exists for the $A \rightarrow Y$ path. In this situation, the regression model estimates none of the causal effects correctly (presuming the total effects are of interest). The estimated effect of X (b_1) is the direct effect of $X \rightarrow Y$ since A is included in the model. And the effect of $A \rightarrow Y$ (b_2) is confounded because of U even though the other confounder (X) is in the model.

Measurement error and DAGs

Chapter 2 presented some discussion of measurement error and how it can result in confounding that cannot be completely removed via conditioning. The problem of measurement error can be readily understood via DAGs. Figure 3.18 illustrates a DAG in which the confounder, Z, is measured with error. According to the CTT model (Crocker & Algina, 1986), it can be decomposed into an unobservable true score component (*Ztrue*) and an observed score component (*Zobs*), both of which are represented in the DAG.

Clearly, conditioning on *Ztrue* would allow unbiased estimation of the causal effect, but this is impossible since the true score cannot be observed. The measured variable *Zobs* can be conditioned on, but itself is not a confounder – it is merely the descendant of one. Conditioning on *Zobs* removes some, but not all, of the *Ztrue*'s variance and therefore provides incomplete blockage of the confounding effect. The variance shared by *Ztrue* and *Zobs* is the reliability coefficient, and the proportion of confounding that can be removed by conditioning on *Zobs* is directly proportional to it.

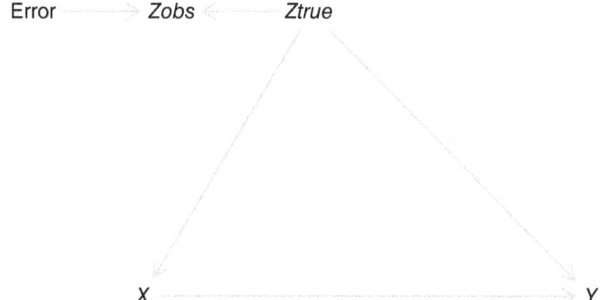

Figure 3.18 A DAG with measurement error. *X* is the exposure variable and *Y* is the response. *Ztrue* is the unobservable confounder true score. *Zobs* is the observed version of the confounder, which is contaminated with measurement error. Conditioning on *Zobs* cannot remove all the *Ztrue*'s confounding

The root cause of measurement error bias is that the true score is a confounder of the relationship between the observed score and the response variable. The researcher estimates the *Zobs* → *Y* relationship when the true DAG is *Zobs* ← *Ztrue* → *Y*. Measurement error bias and confounding bias are fundamentally equivalent.

Using DAGitty

DAGitty is both an R package and a website (www.dagitty.net) that facilitates the drawing, presentation and analysis of DAGs, including deriving their testable implications. The web version is extremely user-friendly and easy to use, while the R package provides additional functionality. Figure 3.19 is a screenshot of the dagitty.net website. In the DAG displayed in this figure, node *E* is the exposure and *D* is the response. *A* is an instrument, *B* is a competing exposure, and *Z* is a collider located between *A* and *B*. In the upper-right corner of the screen, in the area labelled 'causal effect identification', DAGitty reports that no adjustment is necessary to estimate the effect of *D* on *E*. This is because there are no confounders. A simple regression of *D* on *E* would generate an unbiased estimate of the causal effect.

What would happen if the analysis conditioned on *Z*? This can easily be explored in DAGitty. Simply press 'a' on the keyboard while hovering the mouse pointer over *Z* to toggle conditioning. Since *Z* is a collider, conditioning on it opens a confounding path from *E* to *D*. Figure 3.20 shows the DAGitty's analysis of this new scenario.

The causal effect identification pane now reports that there are two sufficient adjustment sets that would allow for the unbiased effect of *E* on *D* to be estimated

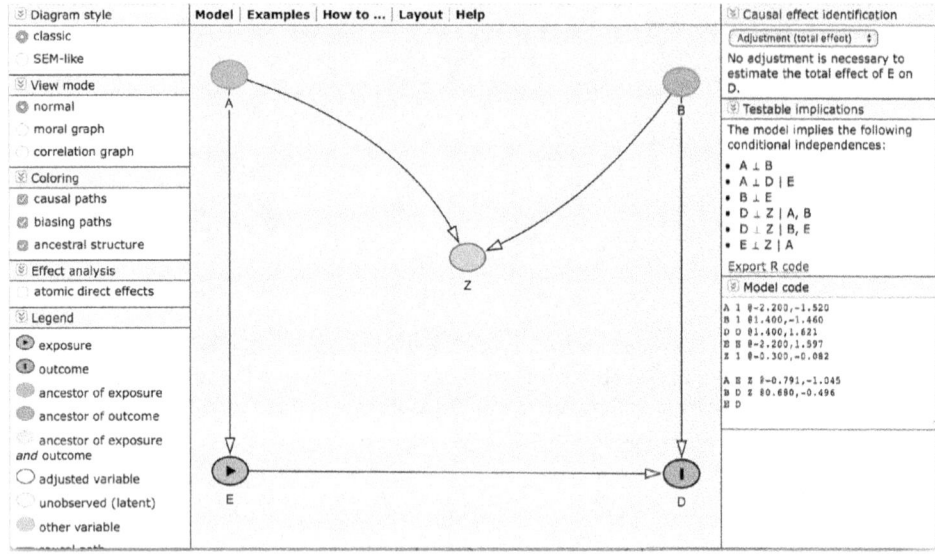

Figure 3.19 dagitty.net is a tool for drawing and analysing DAGs

Note. DAG = directed acyclic graph.

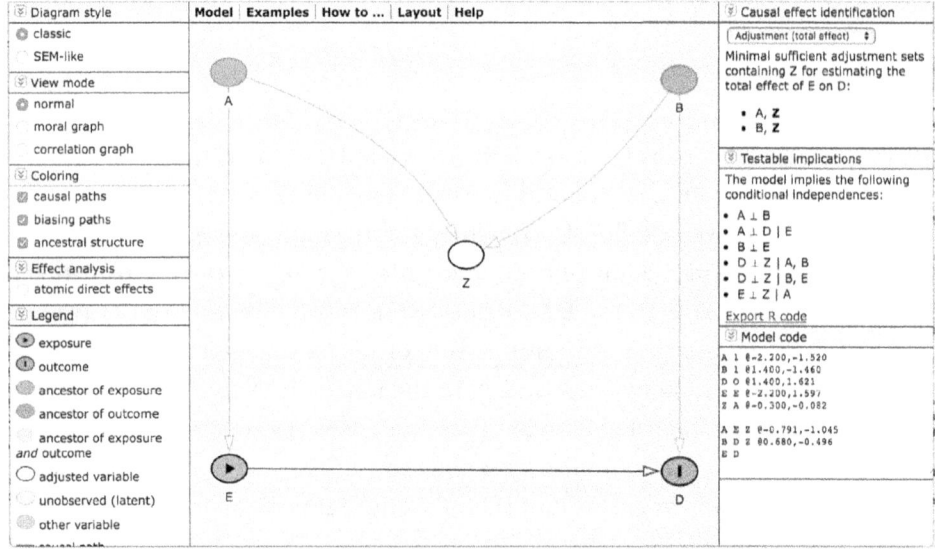

Figure 3.20 dagitty.net analysis of the DAG after conditioning on the collision node Z. Two sufficient adjustment sets are identified

Note. DAG = directed acyclic graph.

even after conditioning on *Z*. They are {*ZA*} and {*ZB*}. Conditioning on either *A* or *B* would block the back-door path opened by conditioning on *Z*. The same output can be produced using the DAGitty R package. The dagitty() function defines the

DAG, and the `exposures()`, `outcomes()` and `adjustedNodes()` functions define the key variables and the conditioning set. The output of the `adjustmentSets()` function, which identifies the minimal sufficient adjustment sets given the graph, is shown below.

```
## { B, Z }
## { A, Z }
```

It is best to not condition on Z at all. But if this is unavoidable, the causal effect can still be recovered by also conditioning on either A or B.

The `impliedConditionalIndependencies()` function reports the DAG's testable implications. Its output is as follows.

```
## A _||_ B
## A _||_ D | E
## B _||_ E
## D _||_ Z | A, B
## D _||_ Z | B, E
## E _||_ Z | A
```

Given the DAG in Figure 3.19, then certain variable pairs are d-separated, or can be rendered d-separated via conditioning and must be independent. For example, $A \perp B$ because they are d-separated by collider Z. And $(A \perp D)|E$ because D is an intermediate node in the chain $A \rightarrow D \rightarrow E$. Therefore, A and D are unconditionally dependent but would become independent after conditioning on E.

The `localTests()` function in DAGitty will automatically test these implications against data. The output of this function reports statistical tests for each implied independence. To pass the tests, each estimate needs to be small enough that its corresponding population value could plausibly be zero. Statistical test result and 95% CIs are provided to assist in this determination. However, keep in mind that statistical tests produce false-positive results (Type I errors) in proportion to the α criterion used, and that the probability of at least one such error (e.g. the *family-wise error rate*) increases rapidly with the number of tests. With six tests, the family-wise error rate is 26.5%, assuming that the tests are independent. The output of the `localTests()` function is given below. The test of independence for $A \perp B$ produced a significant statistical test at the $\alpha = .05$ level.

```
##                      estimate  std.error   p.value       2.5%        97.5%
## A _||_ B            -0.09127326 0.04603154 0.0479341 -0.18171323 -0.000833296
## A _||_ D | E         0.02185533 0.03978672 0.5830379 -0.05631557  0.100026236
## B _||_ E             0.01529169 0.04262281 0.7199209 -0.06845101  0.099034392
## D _||_ Z | A, B      0.03000228 0.05258001 0.5685276 -0.07330473  0.133309293
## D _||_ Z | B, E      0.03662162 0.04984574 0.4628700 -0.06131320  0.134556439
## E _||_ Z | A        -0.04539862 0.04700474 0.3345986 -0.13775112  0.046953870
```

A small proportion of significant tests out of a large family of tested independencies does not necessarily indicate that the graph is definitively wrong. I would recommend adjusting these *p*-values for multiple comparisons before interpreting them using one of the more liberal techniques, such as the Benjamini–Hochberg adjustment (Thissen et al., 2002), which is implemented in R's p.adjust() function. The Benjamini–Hochberg adjusted *p*-value for the test of $A \perp B$ is $p = .288$. After adjustment, this hypothesis test result is no longer particularly concerning.

Testing these implications against data can provide some information about the reasonableness of a hypothesised DAG. Passing these tests is a necessary but not sufficient condition for establishing the utility of a postulated DAG. In other words, if the data does not contain the pattern of independencies implied by the graph, the graph cannot be an accurate description of the data-generating process. But passing the tests does not generally indicate that the DAG is correct. There are likely equivalent DAGs that would imply the same pattern of independences.

The equivalentDAGs() function in DAGitty identifies the equivalent graphs, which are shown in Figure 3.21.

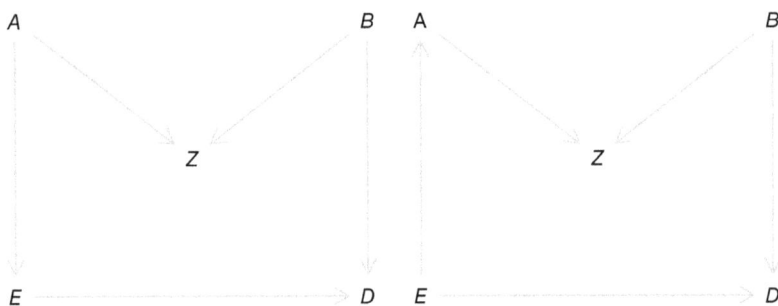

Figure 3.21 Equivalent DAGs. These graphs generate the same implied conditional independencies. They cannot be distinguished on the basis of data

A DAG is a tool for making decisions about conditioning. The existence of equivalent graphs is irrelevant if they all imply the same conditioning sets, and that is the case in this example. Assuming that *Z* is not conditioned on, neither of the equivalent DAGs in the figure requires any adjustment for estimating the causal effect of *E* on *D*. (If *Z* is conditioned on, both DAGs describe either {*ZA*} or {*ZB*} as sufficient adjustment sets.)

Oftentimes some (or all) of the equivalent models can be eliminated on logical or theoretical grounds. For example, an equivalent DAG may describe race as an effect of socio-economic status, an obvious absurdity. When plausible competing DAGs exist which imply differing adjustment sets, this situation deserves to be taken very seriously. Perhaps there exists an adjustment set that is common to both graphs – and conditioning on this particular set renders the analysis robust to the correctness of either DAG. When a common adjustment set does not exist across a set of plausible, equivalent graphs, researchers should consider presenting results from a sensitivity analysis (Saltelli et al., 2004) showing how the resulting causal effect estimates vary across assumed DAGs.

Practical recommendations

The DAG must include all the variables that affect at least two variables in the graph. This means that competing exposures and instruments do not necessarily need to be included. Even so, applying this criterion typically results in a large, complicated diagram. Simple DAGs become easy to analyse visually with practice, but automated tools such as DAGitty are extremely helpful for analysing realistic, complex DAGs. These should be routinely used in study design and data analysis in order to decide which variables to measure and how to estimate the causal effects of interest.

I believe that DAGs are fundamental tools that must be carefully considered when researchers contemplate doing any non-experimental research. A DAG answers three of the most crucial questions that must be answered.

1 Is it even possible to estimate a causal effect, given the available information? If not, the idea must be abandoned. Not all causal research questions can be answered, and it is better to respect unavoidable ignorance than to wish it away.
2 If unbiased estimation is possible, can a conditioning-based method be used? (If not, Chapters 6 and 7 describe alternative approaches that may be possible.)
3 If conditioning can be used, what variables must be included in the adjustment set? What variables must not be included in the adjustment set?

The researcher's working DAG, which serves as the basis for making these fundamental design and analysis decisions, should be prominently disclosed in the manuscript (perhaps as Figure 1). Thus, the researcher's assumptions are clearly communicated to readers, where they may be critically interpreted. If readers find the working DAG to be implausible or unrealistic, then it is likely that they will consider the decisions motivated by it to be flawed and the conclusion derived from it suspect. Either way, the DAG provides a detailed depiction of the researcher's assumptions. Assumptions are necessary to assign meaning to any statistical quantity; they are unavoidable, and they are also often subject to disagreement. The role of DAGs in scientific communication is to highlight the fundamental role of these assumptions and to facilitate meaningful scientific debate about them. It is much more productive to argue about a proposed DAG than to argue about a researcher's seemingly arbitrary selection of method or covariates.

The amount of information needed to create a DAG can seem daunting or even impossible for researchers to possess. It is difficult, but the process is not complete speculation. The ability of a DAG's testable implications to be evaluated against data provides a limited but significant means for identifying and correcting erroneous causal theories. But the difficulty of creating, justifying and defending DAGs does not remove the responsibility for constructing, evaluating and applying them. Decisions regarding how to design the study, what variables to measure, what analysis technique to use and what variables to condition on *must be made*. The question is in the amount of rigor to bring to bear on making those decisions. DAGs provide a principled way to make them. And principled, transparent methods are far superior to the alternative.

This chapter has only scratched the surface of DAGs, on which there is immense literature, and has presented what I believe to be the most immediately useful and relevant concepts for applied researchers. Among other things, DAGs also include a fully specified mathematical framework for describing and calculating the effect of interventions (the *do*-calculus), as well as for reasoning about counterfactuals – those outcomes which were not observed, but could have been (and which are a major consideration of Chapter 4). A full articulation of DAGs is beyond the scope of this book, but the references and recommendations for further reading in this chapter will provide interested readers with a much deeper understanding of this interesting and useful framework.

Further Reading

Elwert F. (2013). Graphical causal models. In S. L. Morgan (Ed.), *Handbook of causal analysis for social research*. Springer. https://doi.org/10.1007/978-94-007-6094-3_13
This book chapter provides an excellent general introduction to DAGs. While more technical than Rohrer's paper, it is clearly written and illustrates how DAGs clarify many concepts in statistics and make it possible for researchers to clearly reason about and solve problems.

Greenland, S., Pearl, J., & Robins, J. M. (1999). Causal diagrams for epidemiological research. *Epidemiology, 10*(1), 37–48. https://doi.org/10.1097/00001648-199901000-00008
This paper is another excellent introduction to DAGs. While it is written for epidemiologists, it should be quite approachable for researchers in other fields as well.

Pearl, J., Glymour, M., & Jewell, N. P. (2016). *Causal inference in statistics: A primer.* Wiley.

This short book provides a relatively non-technical introduction to Pearl's perspective on causal inference in statistics. The key role of information external to the data is emphasised.

Rohrer, J. (2018). Thinking clearly about correlations and causation: Graphical causal models for observational data. *Advances in Methods and Practices in Psychological Science, 1*(1), 27–42. https://doi.org/10.1177/2515245917745629

Rohrer's paper may be the clearest and most readable introduction to the concept of DAGs and their role in study design in the literature to date.

Textor, J., van der Zander, B., Gilthrope, M. S., Liskiewicz, M., & Ellison, G. T. (2016). Robust causal inference using directed acyclic graphs: The R package "dagitty." *International Journal of Epidemiology, 45*(6), 1887–1894. https://doi.org/10.1093/ije/dyw341

This publication introduces and demonstrates the DAGitty software.

Westreich, D., & Greenland, S. (2013). The Table-2 fallacy: Presenting and interpreting confounder and modifier coefficients. *American Journal of Epidemiology, 5*, 292–298. https://doi.org/10.1093/aje/kws412

This paper illustrates some of the insights that are readily available from adopting a DAG perspective by showing that, in general, only one coefficient in a linear model can have a causal interpretation. The general practice of interpreting each row of multiple regression output, as though each has a potentially causal interpretation, is critiqued.

4

RUBIN'S CAUSAL MODEL AND THE PROPENSITY SCORE

Chapter Overview

Rubin's **counterfactual** framework is a theoretical model for defining causal effects and understanding the circumstances under which they can be estimated. This model (like most things in statistics) has many names, including *Rubin's causal model* (RCM), the *Neyman–Rubin counterfactual framework* and the *potential outcomes framework*.

Unlike the Campbell and Stanley (1966) *threats to validity framework*, which was discussed in the introduction to this book, RCM provides a mathematically formalised definition of a causal effect and the conditions under which they can be estimated. Propensity score analysis (PSA) is a method for estimating causal effects which follows directly from Rubin's theory of causal effects. Campbell and Stanley (1966) described an essentially 'negative' description of causation, where causation is the explanation that remains after every threat to validity has been ruled out. Rubin's model provides a positive definition of causal effects: the difference between an outcome occurring under exposure to some treatment versus the counterfactual outcome that would have been observed in the absence of that treatment or under an alternative treatment.

Whereas Campbell and Stanley addressed a wide array of threats to validity, Rubin's approach is similar to Pearl's (2009) DAGs in that the focus is almost exclusively on the confounding threat and its close relatives: the sample selection and mortality threats. Also similar to DAGs, RCM provides a formal description of the conditions under which causation can be assessed. While the terminology and theoretical framework are quite different, both of these approaches end up in the same place regarding whether a causal inference can or cannot be made in a particular situation. A weakness of RCM compared to Pearlian DAGs is that it does not provide much guidance on how to identify which variables need to be conditioned on in order to estimate a causal relationship. The strengths of Rubin's approach are the valuable insights generated by thinking about causal effects in terms of counterfactuals and also in the analytic methods that were inspired by and justified under RCM. These methods can also be justified under the DAGs framework. RCM and DAGs complement each other nicely, and both are certainly worth careful study.

The counterfactual framework

Consider a binary exposure variable D. In the RCM framework, the exposure variable is usually called the *treatment*, although this word is used liberally to describe exposure to conditions that psychologists would not call 'treatments' (e.g. graduating from high school, going to prison, smoking cigarettes, winning the lottery). In the most common notation, the variable D represents exposure to treatment. The subscript i indexes subjects or experimental units.

$D_i = 0$, the ith subject was exposed to control

$D_i = 1$, the ith subject was exposed to treatment

Each subject is either exposed to the treatment or the control condition and is therefore actually observed under only one condition of D.[1] However, according to the RCM framework, each subject has two *potential outcomes* (Y) – one under treatment and one under control. Both of these potential outcomes exist for all subjects regardless of their actual, realised exposure to treatment. According to common RCM notation, these potential outcomes are commonly denoted as Y_0 and Y_1, where

Y_{0i} is the potential outcome for the ith subject in the control condition and

Y_{1i} is the potential outcome for the ith subject in the treatment condition

Defining causal effects under Rubin's causal model

According to RCM, a causal effect is the difference in potential outcomes (Austin, 2011). The causal effect (or treatment effect) for the ith subject is defined as

$$\delta_i = Y_{1i} - Y_{0i} \tag{4.1}$$

The meaning of this equation is more profound than its simple form suggests. Many researchers in the social sciences conceptualise treatment effects in terms of between-subjects comparisons – in other words, comparing different people to each other. The implication of Equation (4.1) is that treatment effects should actually be envisioned as a within-subjects comparison contrasting how an individual would have fared under treatment versus how he or she would have fared under control. But only one of these two potential outcomes is observable. The non-observable outcome is called a counterfactual (Stuart, 2010).

One might think that longitudinal or repeated-measures study designs allow researchers to examine the same participant under the treatment and control conditions, and therefore to observe both potential outcomes, but this is not generally the case. History effects, practice effects, sensitisation or desensitisation and carry-over effects can spoil the second observation, rendering it a poor representation of

---

[1]Though this may seem specific to between-subjects designs, this logic also applies to repeated-measures or within-subjects designs. This is because each subject can only be observed under one particular history of exposures.

the true potential outcome. For some treatments, such as earning a college degree, any comparison of pre- and post-graduation outcomes is confounded by differences in age, maturity and economic conditions. The definition of a treatment effect in Equation (4.1) means that the proper way to conceptualise the causal effect of graduating from college on some outcome (e.g. income) is to compare the income a person would be earning with a college degree at some particular moment in time to the income that the same person would have earned, *at that same moment in time,* without a college degree. To observe both of these potential outcomes would seem to require a time machine or access to parallel universes; otherwise it is impossible. RCM describes the conditions under which one can estimate causal effects using only observable information.

Averaging the treatment effects over all *i* subjects (in other words, taking their *expectation*) yields the average treatment effect (ATE). The **ATE** is the expected value of the causal effect over all of the subjects (Guo & Fraser, 2015). In other words, it is the marginal causal effect. Other types of conditional causal effects will be discussed later in this chapter. (The notation *E()* refers to the expectation operator).

$$\text{ATE} = E(\delta) = E(Y_1 - Y_0) = E(Y_1) - E(Y_0) \tag{4.2}$$

The primary goal of many studies is to arrive at an accurate, precise and unbiased estimate of the ATE or one of the related conditional causal effects that will be described later in this chapter. The ATE represents the expected impact of the treatment on a randomly selected individual from the population of interest. In most cases, treatment effects vary across individuals, but the expected value of the treatment effect is the value that will be least wrong on average (in terms of minimising the sum of squared errors). The ATE is the best single-number summary of the anticipated change in the outcome that will occur when an individual is exposed to treatment, compared with the counterfactual outcome that would have been observed in the absence of treatment (Guo & Fraser, 2015).

The fundamental problem of causal inference

The *fundamental problem of causal inference* is that it is impossible to directly observe the causal effect (Morgan & Winship, 2014). This is because researchers cannot observe the potential outcome under control, Y_{0i}, for all *i* subjects – only those who were actually exposed to the control condition ($D_i = 0$). Similarly, the potential outcome under treatment (Y_{1i}) may only be observed for those who were exposed to treatment ($D_i = 1$). Table 4.1 shows a cross-classification of potential outcomes (*Y*) and treatment status (*D*).

Table 4.1 Observed values and counterfactuals

	Y_0	Y_1
$D = 0$	$Y_0 \mid D = 0$ (observable)	$Y_1 \mid D = 0$ (counterfactual)
$D = 1$	$Y_0 \mid D = 1$ (counterfactual)	$Y_1 \mid D = 1$ (observable)

The conditions on the diagonal of this matrix are observable. The off-diagonal elements are not; they are the *counterfactuals*. One implication of Equation (4.2) is that knowledge of the counterfactuals is required in order to estimate the causal effect of treatment. Why? Because the expected value of the potential outcome under treatment, $E(Y_1)$, is a combination of the observed potential outcome under treatment for the treatment subjects, $E(Y_1 \mid D = 1)$, and the counterfactual potential outcome under treatment for the control subjects, $E(Y_1 \mid D = 0)$. Let π refer to the proportion of the population that is exposed to treatment and $1 - \pi$ the proportion exposed to control. The expected value of the potential outcomes under treatment is then

$$E(Y_1) = \underset{\substack{\text{proportion}\\\text{receiving}\\\text{treatment}}}{(\pi)}\ E(Y_1 \mid D = 1) + \underset{\substack{\text{proportion}\\\text{receiving}\\\text{control}}}{(1 - \pi)}\ E(Y_1 \mid D = 0) \tag{4.3}$$

Correspondingly, the expected value of the potential outcome under control is

$$E(Y_0) = (1 - \pi)\ E(Y_0 \mid D = 0) + (\pi)\ E(Y_0 \mid D = 1) \tag{4.4}$$

It is assumed that members of the population are either assigned to treatment or control, so π and $(1 - \pi)$ must sum to one. Therefore, Equations (4.3) and (4.4) define the potential outcomes as a weighted average of the potential outcome for the treated $(D = 1)$ and the potential outcome for the control $(D = 0)$. When treatment is allocated on a 50/50 basis, the potential outcomes become simple averages of the two conditional potential outcomes.

The meaning of Equations (4.3) and (4.4) can be more easily understood via example. When individuals self-select into treatment, as is often the case outside of randomised experiments, it is likely that those that choose the treatment condition are different from those that choose control on many dimensions. For example, returning to the example of the effect of college completion on income, there are likely to be countless differences on many dimensions between those who complete college versus those who do not (or never enrol). These differences could include sociodemographic characteristics such as race and class; personality variables such as openness to experience, conscientiousness, persistence, academic interests and need for cognition; and cognitive variables such as intelligence. These differences likely extend to, or perhaps even cause, differences on the potential outcomes. If college completers tend to originate from more privileged backgrounds, then it is likely that they would tend to earn more money than non-completers even if they did not complete a college degree. The treatment index variable D represents not just exposure to treatment but also all the other potential systematic differences between the types of people who tend to receive the treatment and those who do not.

Equation (42) describes the causal effect of treatment as the difference in the expected potential outcomes, but each of these expected potential outcomes under treatment and control are themselves mixtures of observable and counterfactual components. Equation (4.5) shows how Equation (4.2) can be rewritten in terms of Equations (4.3) and (4.4), showing explicitly how the ATE depends on the counterfactual potential outcomes (Guo & Fraser, 2015).

$$
ATE = [\underbrace{(\pi)E(Y_1 \mid D = 1)}_{\text{observed}} + \underbrace{(1-\pi)E(Y_1 \mid D = 0)}_{\text{counterfactual}}] -
$$
$$
\underbrace{E(\hat{Y}_1)}
$$
$$
[\underbrace{(1-\pi)E(Y_0 \mid D = 0)}_{\text{observed}} + \underbrace{(\pi)E(Y_0 \mid D = 1)}_{\text{counterfactual}}]
$$
$$
\underbrace{E(\hat{Y}_0)}
$$

(4.5)

Ignorability

Rubin recognised that the fundamental problem of causal inference is, in effect, a missing data problem (Ding & Li, 2018). Therefore, it can be solved by using some of the same techniques he and his colleagues developed for analysing incomplete data. As described in the previous section, the fundamental problem of causal inference is that the treatment effect depends on unobservable quantities – the counterfactual potential outcomes.

Consider an idealised two-group randomised experiment. Such designs are considered the 'gold standard' for estimating causal effects even though the ATE, the estimand of interest, is described by Equation (4.5) and involves counterfactual quantities. These counterfactuals are not problematic in the context of an experiment because of the random assignment of cases to treatment conditions. When the assignment mechanism is randomised, there can be no systematic differences between the treatment units and the control units on any background variables, so $D = 0$ does not represent a different 'type' of person than $D = 1$. The expected value of the difference on any background characteristic of the treated versus the control units is zero. Equivalently, one could state that all background characteristics are independent of treatment assignment under randomisation. The only systematic difference between the treated and control units in an idealised experiment is their exposure to treatment.

This logic also applies to the potential outcomes. When treatment is randomly assigned, the potential outcomes Y_1 and Y_0 are among the infinite set of background characteristics that are equivalent in expectation. This is not to say that $Y_1 = Y_0$ in randomised experiments – that would imply a null effect of treatment. Rather, the expected potential outcomes under treatment or control do not vary across treatment exposure.

Treatment assignment is ignorable when the potential outcomes are independent of treatment (Rubin, 2005).

$$Y_1, Y_0 \perp D \text{ (Ignorability condition)} \tag{4.6}$$

Random assignment of units to treatment conditions creates ignorability. When ignorability holds, then there is no difference in expectation between the observed outcomes under treatment and control and their corresponding counterfactuals.

When ignorability holds,

$$E(Y_1) = E\left[\underbrace{Y_1|D=1}_{\text{observed}}\right] = E\left[\underbrace{Y_1|D=0}_{\text{counterfactual}}\right] \text{ potential outcome under treatment}$$

$$E(Y_0) = E\left[\underbrace{Y_0|D=0}_{\text{observed}}\right] = E\left[\underbrace{Y_0|D=1}_{\text{counterfactual}}\right] \text{ potential outcome under control}$$

$$\tag{4.7}$$

When the ignorability condition holds, the observed outcome under the opposite treatment condition can stand in for or replace the missing counterfactual. It is still true that the counterfactuals cannot be observed, but observing them is no longer necessary because they are equivalent (in expectation) to observable quantities. Under the ignorability condition, the counterfactuals can be ignored. The observed data provide all the information needed to estimate the treatment effect (Guo & Fraser, 2015).

When ignorability holds,

$$ATE = \underbrace{E\left[\underbrace{Y_1|D=1}_{\text{observed}}\right]}_{E(Y1)} - \underbrace{E\left[\underbrace{Y_0|D=0}_{\text{observed}}\right]}_{E(Y0)} \tag{4.8}$$

Ignorability obviously plays a vital role in causal inference in Rubin's model. In the absence of ignorability, it is impossible to generate a valid replacement for the missing counterfactuals, and obtaining an unbiased estimate of the causal effect of the treatment is hopeless. Later in this chapter, the conditions required to achieve ignorability are discussed. There is no definitive statistical or diagnostic test for ignorability. Ignorability is always an assumption that must be justified on logical or theoretical grounds. Estimating a causal effect is therefore a leap of faith, particularly in observational studies – though imperfect compliance or study dropout (mortality) can destroy ignorability even in randomised experiments.

Confounding violates ignorability. It may be helpful to consider 'ignorable' and 'uncounfounded' to be synonyms (VanderWeele & Shpitser, 2011). Ignorability exists when the back-door criterion (see Chapter 3) is satisfied (Pearl, 2014). As discussed at length in previous chapters, conditioning is central to achieving ignorability outside of the context of randomised experiments. But first, let us consider the origin of bias when ignorability does not hold, but researchers nonetheless attempt to estimate a causal effect by comparing the observed outcomes of treatment versus control units.

Bias when ignorability does not exist

Equation (4.8) illustrates how the causal effect of treatment is calculated from the data. It is a simple difference between two observed means (Guo & Fraser, 2015). This equation produces an unbiased estimate of the ATE when ignorability exists due to the equalities described by the equation. But these equalities only exist under ignorability. When ignorability does not hold, Equation (4.8) will produce a biased estimate of the ATE. Performing this calculation is quite straightforward in any statistical software (e.g. the t.test() function in R). A result will be obtained regardless of whether ignorability holds, but in the absence of ignorability, the result will be systematically different from the true value. This type of silent failure is one of the most dangerous hazards of statistical inference.

This section will consider the origin of the bias that results from estimating causal effects in the absence of ignorability. The resulting biased estimate is a mixture of three components: the Actual Treatment Effect (ATE), baseline bias and differential treatment effect bias (Guo & Fraser, 2015).

$$
\begin{aligned}
E(Y_1|D = 1) &- E(Y_0|D = 0) = \\
&E(Y_1 - Y_0) && \text{(the unbiased ATE)} \\
&+ E(Y_0|D = 1) - E(Y_0|D = 0) && \text{(baseline bias)} \\
&+ (\pi)E(\delta|D = 1) - (1-\pi)E(\delta|D = 0) && \text{(differential treatment effect)}
\end{aligned}
\tag{4.9}
$$

where π is the proportion of the sample receiving the treatment and $(1 - \pi)$ is the proportion receiving control.

Baseline Bias

The *baseline bias* component of Equation (4.9) means that the potential outcomes under the control condition differ between those subjects exposed to treatment

($D = 1$) and those exposed to control ($D = 0$; Guo & Fraser, 2015). For example, suppose that the treatment is receiving a college degree, that the outcome is annual income and that a comparison of income for those with college degrees versus those without reveals that degree holders have a higher income. Baseline bias would exist if people with college degrees would have had higher incomes even if they didn't have degrees, perhaps by virtue of differences in socio-economic status, prior school achievement, intelligence, motivation, personality or other background factors.

Differential Treatment Effect Bias

The *differential treatment effects* component of Equation (4.9) means that the two groups of subjects ($D = 0$ and $D = 1$) may experience different benefits of the treatment (Guo & Fraser, 2015). For example, intellectually curious and highly motivated people experience higher returns from educational experiences than do incurious or unmotivated people. Even when those people who choose to attend college ($D = 1$) experience large treatment effects (e.g. ($Y_1|D = 1$) – ($Y_0|D = 1$) is large), it does not necessarily follow that the people who chose not to attend college would have experienced the same benefits. This is especially true when individuals self-select into 'treatments' on the basis of their anticipated returns from those treatments.

Conditional ignorability

So far, only one situation has been described that could plausibly create ignorability (Equation 4.6) which is the random assignment of subjects to treatment conditions. It is possible to achieve ignorability outside of the context of randomised experiments by identifying a set of confounder variables, $\mathbf{X}$, which are causally related to both selection into treatment as well as the potential outcomes. Conditioning on these variables renders the potential outcomes Y_0, Y_1 conditionally independent of treatment assignment, D (Austin, 2011).

$$Y_1, Y_0 \perp D|\mathbf{X} \qquad \text{(Ignorability conditional on } \mathbf{X}) \qquad (4.10)$$

The conditional ignorability condition described by Equation (4.10) is equivalent to d-separation as described in Chapter 3, except that it is expressed in terms of potential outcomes rather than the observed outcome. The potential outcomes are theoretical quantities that are a mixture of observed and counterfactual components. But the central idea is the same – conditioning on an appropriate set of confounding variables allows for the unbiased estimation of causal effects. In the DAG framework, this conditioning is viewed as blocking all the back-door paths between the exposure variable and the outcome. In Rubin's counterfactual frame-work, the observed and counterfactual components of each potential outcome become the same after conditioning, so the missing counterfactual components become irrelevant. These are two different theoretical perspectives on the same basic phenomenon.

Conditional treatment effects

The ATE is a marginal treatment effect that averages over the entire population. The possibility of differential treatment effects over levels of D means that two additional types of conditional treatment effects can be considered: the average treatment effect for the treated (ATT) and the average treatment effect for the untreated (ATU; Guo & Fraser, 2015).

The **ATT** is the treatment effect for those subjects that received the treatment. It is the difference in potential outcomes specifically for the cases that received the treatment.

$$\text{ATT} = E(Y_1|D = 1) - E(Y_0|D = 1) \tag{4.11}$$

The **ATU** is the treatment effect for those subjects that received the control. Alterna-tively, this is sometimes called the *average treatment effect for control* (ATC). It is the difference in potential outcomes for the cases that did not receive the treatment.

$$\text{ATU} = E(Y_1|D = 0) - E(Y_0|D = 0) \tag{4.12}$$

If ATT ≠ ATU, then there are differential treatment effects, implying that D is a mod-erator of the treatment effect (Morgan & Winship, 2014). The ATE is a weighted average of ATT and ATU, with the weight for ATT dependent on the proportion of the subjects who received the treatment (π) and the weight for ATU on the proportion of subjects who did not receive the treatment ($1 - \pi$).

$$\text{ATE} = (\pi)\text{ATT} + (1 - \pi)\text{ATU} \tag{4.13}$$

These conditional treatment effects can be extremely relevant for answering policy questions, sometimes even more than the ATE. For example, the ATT answers the following questions:

- How much does smoking harm people *who choose to smoke*?
- What health effects do *distance runners* experience from running?
- How does marriage affect the long-term happiness of *married people*?

While the ATU answers the following questions:

- If *high school dropouts* had graduated, how would their rate of labour market participation change?
- How severely would *non-smokers* be harmed if they started smoking?

Social research is often motivated by an underlying desire to improve outcomes for people who are perceived to be suffering. The struggles of high school dropouts in terms of employment, income and poverty status are well-documented. It is natural to ask how these people could be helped. If sufficient support could enable them to achieve a high school diploma or equivalent, how would their lives change? The ATU is far more relevant to answering this question than the ATE because it applies specifically to dropouts (e.g. $D = 0$, if high school completion is viewed as the treatment), whereas the ATE applies the 'average' member of the population, which is overwhelmingly composed of high school graduates.

One interesting detail about the ATT and ATU treatment effects is that, unlike ATE, they cannot generally be estimated from randomised experiments. Randomisation implies that all background characteristics, including the potential outcomes, are equivalent in expectation between the treatment and control groups, implying that the treatment effect for $D = 0$ must be identical to the treatment effect for $D = 1$ – which means that, under randomisation, ATE, ATT and ATU all have the same value (Sobel, 2009).

Although non-randomised quasi-experimental or correlational studies are much less reliable because of their exposure to confounding bias, they do allow for the estimation of ATT and ATU, which often cuts to the heart of what researchers actually hope to learn when they engage in research. The limited inferential possibility resulting from random assignment to treatment is a shortcoming that is rarely noted. In exchange for this limitation, randomisation offers freedom from the many stringent and difficult to satisfy assumptions about the identity, proper measurement and effectively modelling of the confounders.

Example: estimating ATT, ATU and ATE via linear regression

The distinction between ATT and ATU can be more easily understood if one examines a relatively simple situation in which D is a binary indicator of treatment, the set of covariates **X** includes only a single variable X and linearity and normality assumptions are warranted. In that case, all three treatment effects (ATE, ATT and ATU) can be estimated by a linear regression model of the following form:

$$Y_i = b_0 + b_1(D_i) + b_2(X_i) + b_3(D_i X_i) + e_i \qquad (4.14)$$

Note that the inclusion of parameter b_3 allows the treatment effect to be moderated by X. It is this moderation that creates the differing values of ATE, ATT and ATU. The three treatment effects can be estimated by calculating conditional slopes for the treatment indicator ('simple slopes') after centring X at different values. This process is sometimes called 'probing interactions' in regression literature (Bauer & Curran, 2005).

For example, assume that X is a composite variable scored on a standardised (z-score) metric representing health behaviours, perhaps inclusive of BMI (body mass index), alcohol intake, adequate sleep, nutrition and exercise. Let the treatment variable D represent smoking status, and the response variable Y represent lung function, also measured on an age-adjusted standardised metric.

Assume that the values presented in Table 4.2 describe the health behaviour (X) means for smokers, non-smokers and the grand (marginal) mean.

Table 4.2   Health behaviour composite means by group

	Health behaviours mean	Proportion of population
Smokers ($D = 1$)	$\bar{X} = -0.65$	20%
Non-smokers ($D = 0$)	$\bar{X} = 0.2$	80%
Overall (marginal)	$\bar{X} = 0.3$	

Further, assume that the regression model described in Equation (4.14) is fit to the data, producing the following results.

```
##
## Call:
## lm(formula = Y ~ D * X)
##
## Residuals:
##     Min      1Q  Median      3Q     Max
## -6.1082 -0.9904  0.0342  0.9916  5.1449
##
## Coefficients:
##               Estimate Std. Error t value Pr(>|t|)
## (Intercept)    0.05223    0.05463   0.956    0.339
## D             -0.79688    0.13635  -5.844 6.89e-09 ***
## X              0.71522    0.05186  13.791  < 2e-16 ***
## D:X            0.58158    0.11677   4.981 7.47e-07 ***
## ---
## Signif. codes:  0 '***' 0.001 '**' 0.01 '*' 0.05 '.' 0.1 ' ' 1
##
## Residual standard error: 1.49 on 996 degrees of freedom
## Multiple R-squared:  0.2577, Adjusted R-squared:  0.2555
## F-statistic: 115.3 on 3 and 996 DF,  p-value: < 2.2e-16
```

The conditional treatment effects are recovered as follows:

ATT: Set $X = -.65$. The simple slope for D is then $\underbrace{-0.797}_{b_1} + \underbrace{(-.65)}_{\bar{X}|D=1}\underbrace{(0.582)}_{b_3} = -1.175$.

ATU: Set $X = .20$. The simple slope for D is then $\underbrace{-0.797}_{b_1} + \underbrace{(.20)}_{\bar{X}|D=0}\underbrace{(0.582)}_{b_3} = -0.681$.

Comparing these two estimates reveals that the negative health effects of smoking are more severe for smokers than they would be for non-smokers (if non-smokers were 'treated' with smoking). This is a consequence of their lower health behaviours scores (X), revealing that smokers tend to sleep and exercise less, eat worse and drink more. Perhaps these health behaviours buffer the effects of smoking such that smoking is less harmful for people leading otherwise healthy lifestyles.

The ATE can be estimated by centring X to the marginal mean and calculating the simple slope of D.

ATE. Set $X = .03$. The simple slope for D is then $-0.797 + (.03)\ (0.582) = -0.78$.

Equation (4.13) shows how the ATE can be alternatively estimated from ATT and ATU.

$$
\begin{aligned}
\text{ATE} &= (\pi)\text{ATT} + (1 - \pi)\text{ATU} \\
&= (.2)(-1.175) + (1 - .2)(-0.681) \\
&= -0.78
\end{aligned}
$$

This value is identical to the one obtained by calculating ATE via the simple slopes approach.

The propensity score

The previous section showed how conditioning on an appropriate set of background variables X allows one to achieve conditional independence and produce unbiased estimates of causal effects even in the absence of randomisation. This was further illustrated by a trivial example, based on simulated data, in which X consisted of only a single variable and the linear regression assumptions were fulfilled.

In many realistic situations, X contains a large set of variables. As described in Chapter 2, there are several ways to implement conditioning: statistical control, sampling, subclassification, matching and weighting. Of these, the only method that can gracefully accommodate large numbers of conditioning variables is statistical control. The other methods suffer from the curse of dimensionality; they become exponentially more intractable as the number of variables increases. But statistical control has serious drawbacks – its performance relies on the satisfaction of a large set of assumptions regarding model specification, and it will happily produce estimated causal effects even when there is no overlap between the treated and control cases on the variables in **X**. In this case, the results are exceptionally assumption leveraged and rather poorly informed by the data.

There is a solution to the curse of dimensionality. Rosenbaum and Rubin (1983) showed that conditional ignorability can be achieved by conditioning on the propensity score – the conditional probability of being treated given **X**. The high-dimensional variable set **X** can be replaced by a single vector of propensity scores. The propensity scores can then be conditioned on using any method, including non-parametric methods such as subclassification, matching or weighting that are leveraged on fewer assumptions than statistical control. The propensity score can be written as follows:

$$p(D = 1|\mathbf{X}) \qquad\qquad \text{(Propensity score)} \qquad\qquad (4.15)$$

If conditioning on a set of covariates **X** is sufficient for ignorability, conditioning on the propensity scores based on **X** is also sufficient for achieving ignorability (Rosenbaum & Rubin, 1983).

$$Y_1, Y_0 \perp p(D = 1|\mathbf{X}) \qquad \text{(Ignorability conditional on the propensity score)} \qquad (4.16)$$

Approximating an experiment

It is instructive to contrast the equivalence between groups that results from random assignment with the equivalence that is a result of conditioning. Conditioning methods force all cases to have a particular value (or nearly so) of one or more background variables. For example, imagine that a researcher wants to condition on IQ via sampling. This researcher could recruit a large number of potential subjects, measuring each one's IQ, and retain only those whose IQ scores are approximately 100. Any analysis run on such a sample has implicitly conditioned on IQ. The same obviously occurs when researchers use subclassification as the conditioning method – within strata, subjects are closely equivalent on the variable. This also applies to matching, as subjects within matched sets are very similar on the matching variable. And it even applies to statistical control, which projects the effect of each variable into an orthogonal space, allowing the effect of each variable to be computed as though subjects were strictly equivalent on the other variables. Thus, all conditioning strategies result in exact equivalence – or as near as can practically be achieved – on a specific set of background variables.

Random assignment does not result in this type of exact equivalence (Deaton & Cartwright, 2018). If subjects are randomly assigned to groups, background variables (e.g. IQ) will vary – perhaps widely – within those groups. Randomisation results in equivalence in distribution, not equivalence in value. In the long run, random assignment would result in an identical distribution of IQ across groups. This distributional equivalence is 'good enough' for ignorability even though it is a weaker condition than the exact equivalence that can be obtained from conditioning. This condition is called *equivalence in expectation* or *balance in expectation*. The great advantage of randomisation is that it produces equivalence in expectation for all background variables, both observed and unobserved, including the potential outcomes. It does this without forcing researchers to identify, precisely measure and condition on those variables.

Conditioning on the propensity score produces equivalence in expectation, not equivalence in value. Assuming that **X** contains more than one variable, it is typically the case that a particular value of the propensity score could be obtained by many different combinations of values of the variables in **X**.

For example, assume that **X** contains two continuous variables, X_1 and X_2, both of which have positive relationships with the probability of receiving treatment (and therefore, the propensity score). Further, assume that the propensity score (p) is related to X_1 and X_2 via the following logistic regression equation (Menard, 2002).

$$\log\left(\frac{p}{1-p}\right) = .25 + .7(X_1) + .5(X_2)$$

Now, choose a particular value of the propensity score. I will arbitrarily choose $p = .8$. Subjects with such a propensity score have an 80% chance of being treated given their values of X_1 and X_2. Figure 4.1 shows the combinations of values for X_1 and X_2 that result in a propensity score of $p = .8$. (This is a straight line because of the additivity and linearity of the logistic regression model that I used to generate the data.) The point is that any constant choice of propensity score can be generated using many combinations of values of the covariates. Fixing the propensity score, which is essentially what happens then conditioning on it, does not 'pin down' values of the involved covariates to specific values, but does impose a fixed trade-off among these variables.

Figure 4.2 provides a graphical representation of what *equivalent in expectation* means. This figure is based on simulated data. I selected all the cases with propensity scores equal to $p = .8$ and plotted the densities of X_1 and X_2 by treatment status on the left and right panels of the figure. The distributions of these two variables exhibit near-perfect equivalence in distribution across treatment conditions after conditioning on the propensity score. Any remaining imbalance is a result of sampling error

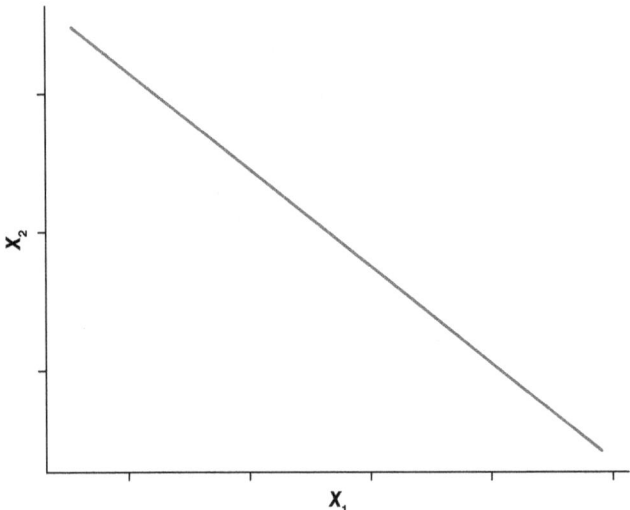

Figure 4.1 Combinations of the values of X_1 and X_2 that produce a propensity score of $p = .8$

due to the finite sample that was used. It is obvious that X_1 and X_2 were not forced to take on some specific values that are equal across treatment groups – there remains a good deal of variability in these two variables after conditioning on the propensity score. But the distributions are balanced, particularly with respect to their expected values (e.g. the means).

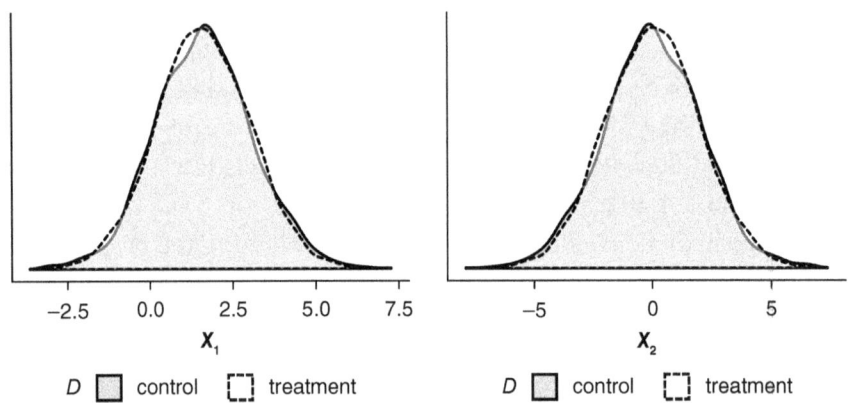

Figure 4.2 Equivalence in distribution on X variables after conditioning on the propensity score

For X_1, the means for the treatment and control groups are 1.609 and 1.626, respectively. For X_2, the means are 0.02 and –0.003.

Simulated data set

The examples of propensity score estimation in this chapter will be based on data simulated from the DAG shown in Figure 4.3. Continuing the previous notation, the exposure variable D is a binary treatment variable and Y is a continuous response variable. Variables A and B are confounders, which cause both the exposure/treatment variable D as well as the outcome Y. The influence of the confounding variables creates an additional statistical relationship between D and Y that needs to be removed in order to render this empirical relationship an unbiased representation of the *causal* influence of D on Y.

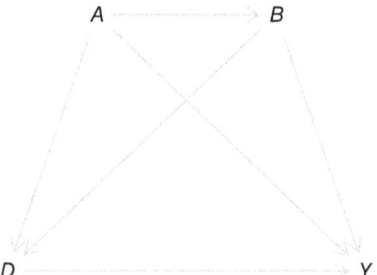

Figure 4.3 Example of a DAG for the propensity score analysis examples

Note. DAG = directed acyclic graph.

The synthetic data set generated from this DAG contains $n = 700$ records and can be downloaded from the book's website.

Propensity scores

Propensity scores represent each subject or unit's model-predicted probability of being in the treatment condition given a set of background variables, **X**. The variables in **X** are those that are capable of creating conditional independence of the potential outcomes from treatment assignment. This implies that **X** must contain a *sufficient adjustment set* that blocks all confounding paths, but it must not contain any mediators or colliders. Working with a DAG can be extremely helpful in identifying what variables to include in the propensity score model. In practice, researchers often attempt to articulate (and measure) the full set of variables that they believe influence selection into treatment. If done successfully, this will result in conditional independence, even though it may result in including more variables in the propensity score model than are strictly necessary. Of course, as Chapter 2 discussed in detail, all variables must be measured with high validity in order for any conditioning strategy to succeed.

The identification and measurement of the variables in **X** is the most critical issue in PSA – or indeed, any causal inference technique that is based on conditioning. The ability of propensity score techniques to achieve unbiased estimates of causal effects depends entirely on the degree to which researchers are able to assemble a set of variables that can create conditional independence. If this is not achieved, no degree of statistical trickery or algorithmic sophistication will enable unbiased results to be obtained. The identification of these variables is, without a doubt, the central issue in PSA. All other concerns are mere details by comparison.

The propensity scores themselves can be estimated in a variety of ways. Any statistical technique or classification algorithm that can calculate the predicted probability of class membership given **X** can be used to compute propensity scores. The two most dominant approaches, which will be discussed in this chapter, are logistic regression and boosted classification trees, but many other methods (e.g. support vector machines, random forests or probit regression) could be used. The most important distinction between the methods for computing propensity scores is whether researchers must articulate a full statistical model of the relationship between the **X** variables and the probability of treatment or whether the model itself is learnt from the data. This categorical distinction separates methods based on statistical models from newer techniques based on machine learning (James et al., 2013). Logistic regression and boosted classification trees are prototypical examples of each of these.

Estimating propensity scores via logistic regression

Logistic regression is an extension of linear regression to binary categorical outcomes (Menard, 2002). It is a member of the generalised linear model family with a logit link and a Bernoulli distribution for the response variable. The general form of a logistic regression model is written as

$$\log\left(\frac{p_i}{1-p_i}\right) = b_0 + b_1 X_{1i} + b_2 X_{2i} + \cdots + b_k X_{ki} \tag{4.17}$$

where p is the probability of the response variable having a particular categorical membership (e.g. $Y = 1$ as opposed to $Y = 0$), i indexes experimental units or subjects and k indexes the total number of explanatory variables.

The *logit link function*, $\log\left(\frac{p_i}{1-p_i}\right)$, maps the [0,1] range of probabilities to the $(-\infty, \infty)$ range of values that may be produced by the right-hand side of the model – the *linear predictor*. The logit link produces an s-shaped or sigmoid function relating values of the linear predictor to predicted probabilities (p). Figure 4.4 illustrates this mapping.

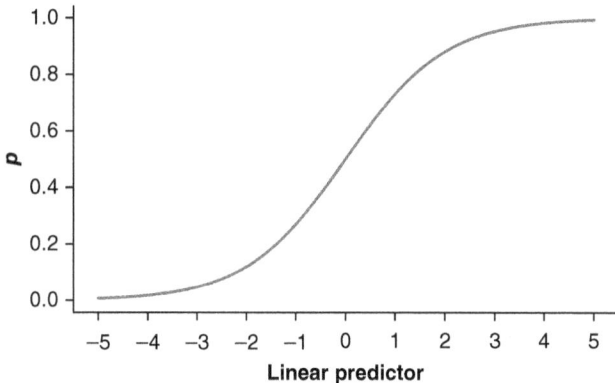

Figure 4.4 Logit link mapping values of the linear predictor to p

Computing or estimating propensity scores with logistic regression is simple. The steps are as follows:

1   Fit a logistic regression model to the data with treatment status (D) as the response variable and the set of variables deemed necessary to create conditional ignorability (**X**) as the predictors.
2   Use the resulting model to calculate, for each case, the model-implied (or 'predicted') probability of receiving treatment given that case's values on the **X** variables. These are the propensity scores.
3   Append the propensity scores to the data set for later use.

Note that the model in step 1 must be complete and correct. If treatment status is influenced by the interaction of variables X_1 and X_2, then this interaction term must be included in the logistic regression model. The same is true of non-linear relationships between variables and the probability of response. Many analysts routinely limit the logistic regression model for the propensity scores to linear main effects only. This is a mistake. This model, like all models, must be thoughtfully constructed. Careful attention must be devoted to the plausibility of model assumptions. Indeed, this requirement is one reason that many analysts have moved to machine learning techniques for estimating the propensity scores, as they prefer to be relieved of the responsibility for correct (or at least reasonable) model specification.

In R, the `glm()` function can be used for fitting logistic regression models. Once the model is fit, the `predict()` function is used to calculate the propensity scores from the model.

Fitting a logistic regression model in R with main effects of A and B plus their inter-action AB produces the following output.

```
## 
## Call:
## glm(formula = D ~ A * B, family = "binomial", data = data, na.action =
## na.exclude)
## 
## Deviance Residuals:
##      Min       1Q    Median       3Q      Max
## -2.1290  -0.8711   -0.4279   0.8962   2.4779
## 
## Coefficients:
##                  Estimate Std. Error z value Pr(>|z|)
## (Intercept) -0.261438    0.098117  -2.665  0.00771 **
## A            1.402105    0.122684  11.429  < 2e-16 ***
## B            0.939218    0.107218   8.760  < 2e-16 ***
## A:B          0.003771    0.077543   0.049  0.96121
## ---
## Signif. codes:  0 '***' 0.001 '**' 0.01 '*' 0.05 '.' 0.1 ' ' 1
## 
## (Dispersion parameter for binomial family taken to be 1)
## 
##      Null deviance: 962.14  on 699  degrees of freedom
## Residual deviance: 759.99  on 696  degrees of freedom
## AIC: 767.99
## 
## Number of Fisher Scoring iterations: 4
```

The output reveals that both A and B have strong and statistically significant relation-ships with the probability of being treated, whereas the effect of their interaction is nearly zero.

Figure 4.5 displays the densities of the propensity scores by treatment status. The figure evinces two features that must be present in order for PSA to function properly. First, there is overlap in the propensity score distribution across groups – a feature known as **common support** (Guo & Fraser, 2015). All methods of PSA require that at least some subset of the cases have similar propensity scores across treatment con-ditions. Without common support, it would prove to be impossible to closely match cases across treatment conditions with similar propensity scores, as no such cases exist. Nor would subclassification be possible, because no stratum of reasonable size could contain both treated and untreated cases.

Second, the propensity score distributions are distinguishable with differing modal values. According to Figure 4.5, the modal propensity score for non-treatment cases

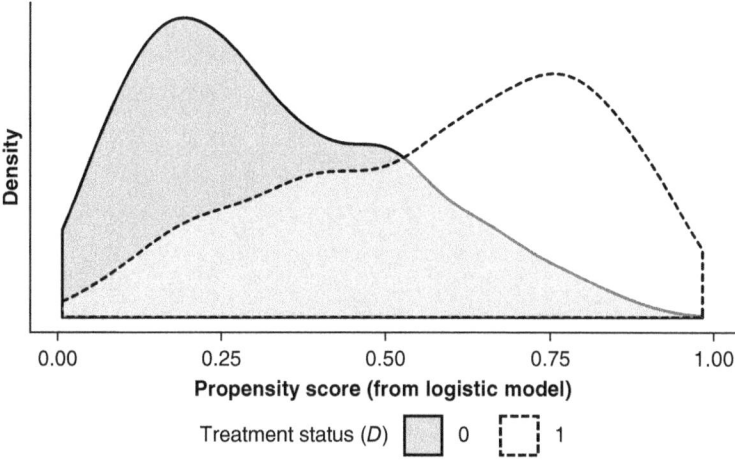

Figure 4.5 Logistic regression–derived propensity score density by treatment status from the analysis data

is about $p = .2$, while the modal value for the treated is about $p = .8$. The distinguishability of these distributions provides evidence that the **X** variables, in aggregate, do indeed discriminate to some degree between the treated and the untreated. (More formally, that the *model* constructed from the **X** variables, as specified by the analyst, discriminates between groups). When this condition is absent, it suggests that either the set of predictor variables included in the propensity score is incorrect or that perhaps the natural assignment mechanism is nondeterministic or stochastic. In either case, propensity score methods are unwarranted or possibly even unnecessary. If the wrong variables were selected, then there is no hope of achieving ignorability by conditioning on the propensity scores. If the true assignment mechanism is stochastic, no conditioning method is required to reach an unbiased estimate of the causal effect because nature has provided randomisation.

Figure 4.6 shows an example of complete separation on the propensity scores across treatment conditions. Complete separation can result from overfitting the propensity score model or can simply reflect the reality of the social determinants of treatment. It is possible for the treated and untreated subjects to be entirely distinct with respect to their background characteristics. In such a case, conditioning on the propensity score cannot approximate an experiment.

A more realistic situation is partial overlap of the propensity score distributions for the treatment and control groups, as illustrated by Figure 4.7. In this case, the analysis should drop (or trim) cases from the non-overlapping regions of the distribution and proceed with the analysis (Guo & Fraser, 2015). This discarding of data may jeopardise the external validity of the study, because the subpopulation of cases with similar propensity scores may not resemble the broader population of interest.

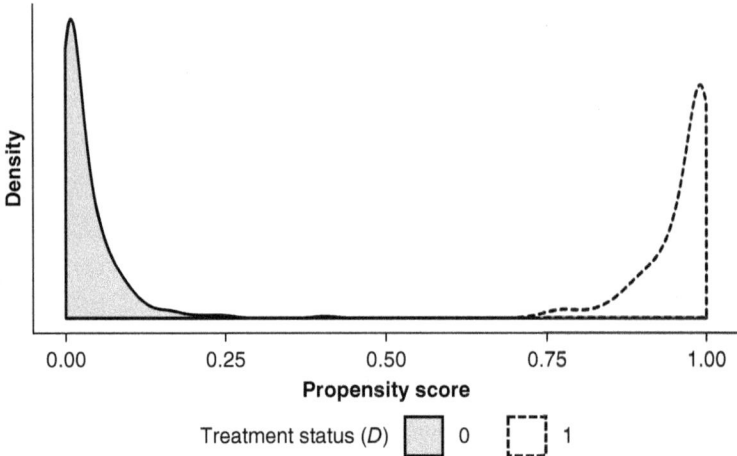

Figure 4.6 Example of propensity score distribution with no common support. Causal effects cannot be estimated with propensity score methods under this condition

Further, discarding data reduces precision and statistical power. But there is no alternative unless researchers are willing to employ covariate adjustment and extrapolate across the empty regions. Crump et al. (2009) provided a principled method for determining which cases to discard in this circumstance.

Another diagnostic that should be checked is the classification accuracy of the propensity scores estimated via the logistic regression model. One way to obtain this information is to create a 'confusion matrix': a 2 × 2 contingency table of model-predicted treatment status versus actual treatment status (Table 4.3). I created a binary predicted treatment variable from the propensity scores by rounding them to zero decimal places using the `round()` function, which will produce a predicted treatment value of one for all cases with $p > .5$ and a predicted treatment value of zero otherwise. The contingency table was created using the `table()` function.

Table 4.3 Logistic regression-derived propensity score-predicted treatment status versus actual treatment status

	Actual D = 0	Actual D = 1
Predicted D = 0	297	107
Predicted D = 1	91	205

The values on the diagonal are the correct classifications; the off-diagonal values are the incorrect classifications, so this table indicates 502 hits and 198 misses. The correct classification rate is the sum of the diagonal values divided by the total number of cases; in this case, the rate is approximately 71.7%.

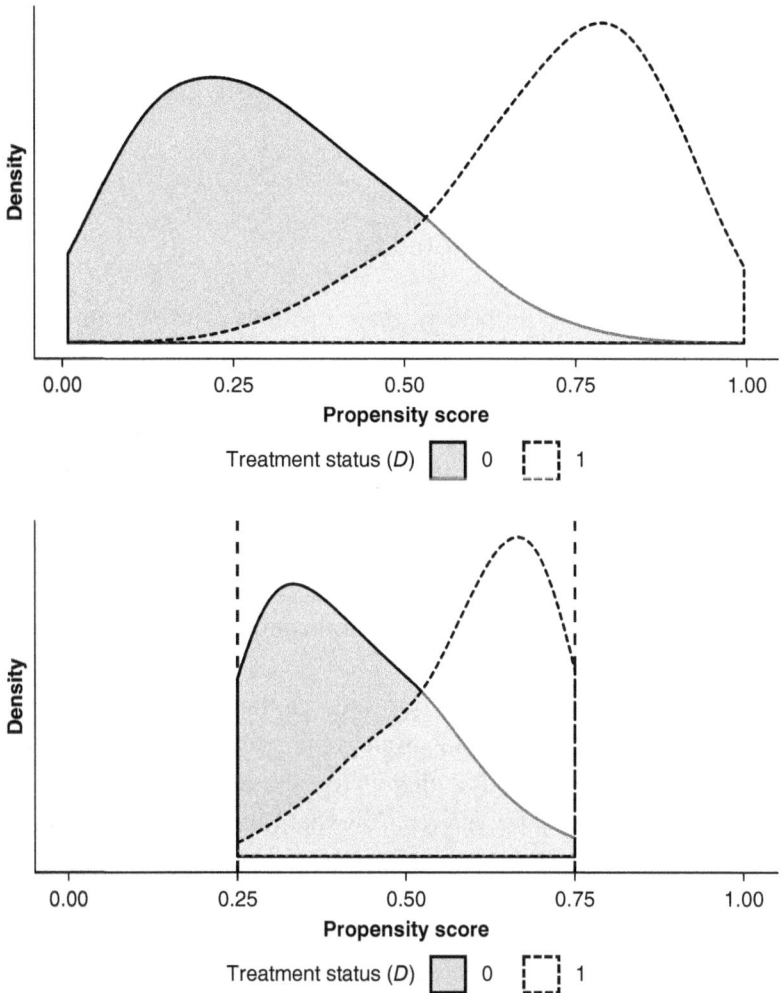

Figure 4.7 Example of propensity score distribution with moderate overlap. Left panel: Distribution of propensity scores by group. Right panel: Distribution of propensity scores after discarding cases

The expected classification rate in the absence of an informed model is not zero but rather the maximum of π or $(1 - \pi)$, where π is the proportion of treated cases (Menard, 2002). For example, if the sample is 30% treatment and 70% control, one could obtain 70% classification accuracy by simply assigning all cases to the control group. In our example data, 44.6% of the subjects received treatment, so the baseline classification accuracy is 55.4%. The classification rate of 71.7% looks less impressive compared to this standard, but very high hit rates often indicate a lack of common support. In my experience, classification hit rates in the 75% to 85% range are achievable and satisfactory, but this certainly must vary across research contexts. The important question is whether conditioning on the propensity scores balances the sample on all of the relevant variables – which is something that can easily be determined regardless of which conditioning method the analyst chooses. Details on how to use the propensity scores to estimate causal effects will be presented in Chapter 5.

Solving the curse of dimensionality

The primary motivation of propensity score methods is in solving the 'curse of dimensionality', enabling the use of conditioning methods such as weighting or subclassification that can only be realistically applied to a single variable. Matching can be performed in high-dimensional spaces on the basis of *Mahalanobis distance*, a measure of the distance between points in space that accounts for correlations between variables (Gu & Rosenbaum, 1993), but nominal and ordinal variables are a challenge for this metric. (For example, how does one measure the 'distance' between genders or races?) It is often more straightforward to match on the one-dimensional propensity scores and accept the resulting weaker condition of *matching in expectation* rather than *matching in value*.

Figure 4.8 shows how propensity scores collapse the multivariate space of the **X** variables onto a single axis. Figure 4.9 displays the resulting one-dimensional distribution of propensity scores. Conditioning on (or adjusting for) the propensity scores is sufficient for estimating causal effects. Note that the y-axis values have been randomly disturbed ('jittered') in this figure to prevent overplotting.

Propensity score estimation via boosted classification trees

The boosted classification tree is a relatively new machine learning technique for supervised classification. The term *supervised* in this context means that the algorithm is initially trained on labelled data in which the class membership is known. Although a full overview of this technique is beyond the scope of this chapter, I hope a brief introduction is useful.

A **classification tree** is a non-parametric technique for classifying data (Breiman et al., 1984). The method is based on successive partitioning of the sample into subsets that maximise within-subset similarity on the categorical response variable. Figure 4.10 displays an example classification tree calculated from the simulated data set. Beginning at the top of the figure, the first partition splits the data at the value $A = .046$. The subset of data with $A < .046$ is on the left branch, the subset with $A \geq .046$ is on the right. In general, the left branches in the diagram indicate that the test condition is true; right branches indicate that it is not true.

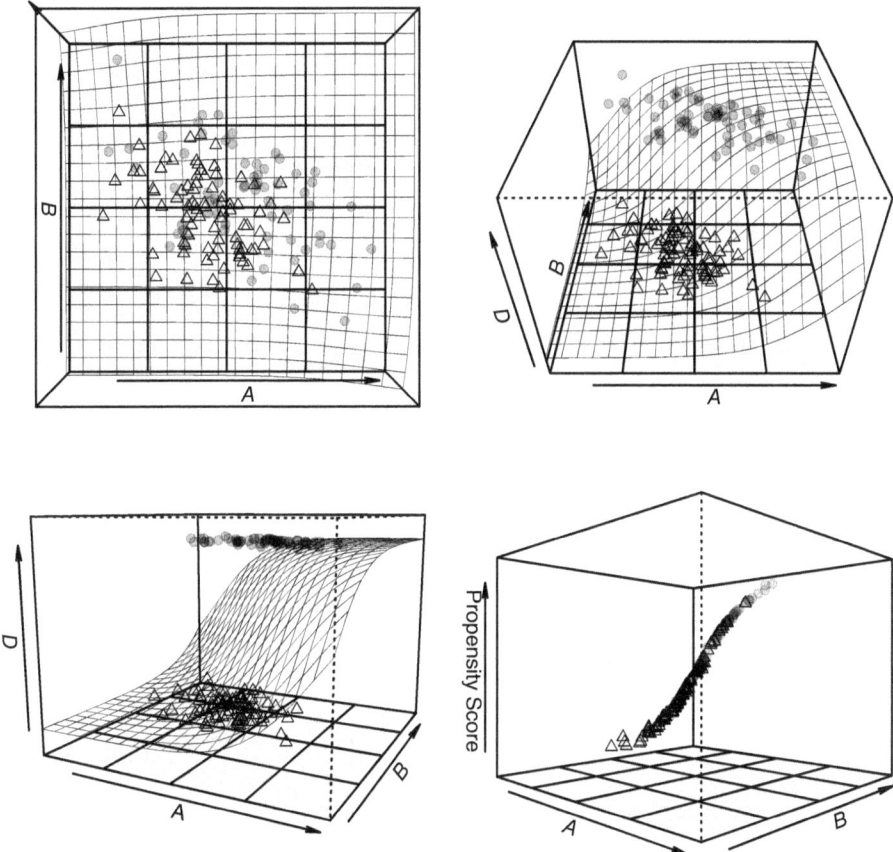

Figure 4.8 Propensity score dimension reduction. Black circles are treatment cases. White triangles are control cases. Panel 1: Three-dimensional scatter plot of confounders *A* (*x*-axis) and *B* (*y*-axis) versus treatment status (*D*; *z*-axis + shape and colour) viewed from above. Panel 2: Rotation to show the points plotted against the predicted probability (propensity score) surface. Panel 3: Further rotation to display the sigmoid shape of the surface with vertical reference lines connecting each point to its estimated propensity score. Panel 4: Location of each point with respect to the propensity score (*z*-axis). The formerly two-dimensional location of each point has been reduced to its location along a one-dimensional number line

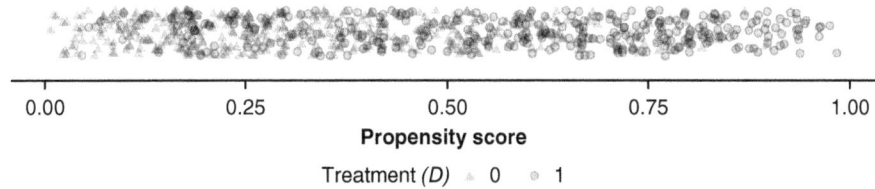

| 0.00 | 0.25 | 0.50 | 0.75 | 1.00 |

Propensity score

Treatment *(D)* ▲ 0 ⊙ 1

Figure 4.9 The two-dimensional space of variables *A* and *B* is reduced to a one-dimensional location along a number line of propensity scores

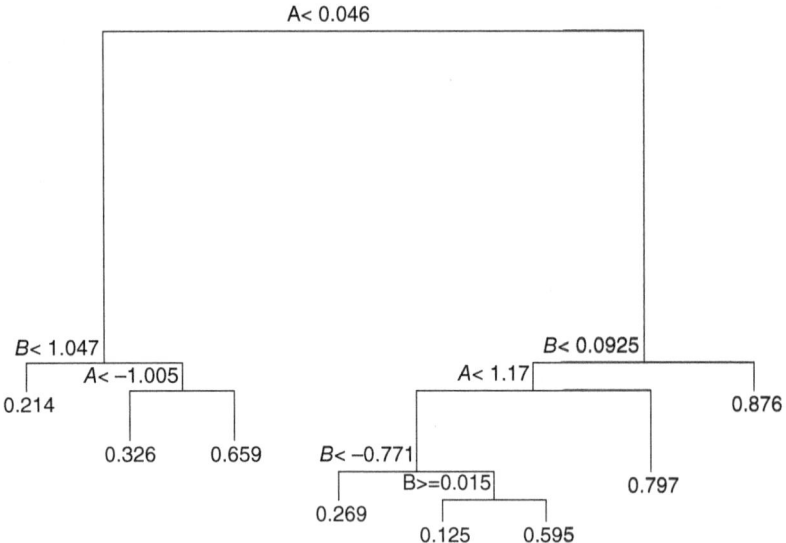

Figure 4.10 Example of classification tree. The numbers below each leaf are the predicted probabilities of treatment

Continuing down the tree on the left, the next division is at $B = 1.047$. Cases for whom $A < .046$ and $B < 1.047$ have a predicted probability of treatment of 0.214. The termination point of a classification tree is called a *leaf*. Each leaf is associated with a predicted probability of class membership (Moisen, 2008).

An additional partition was made for those cases with $B \geq 1.047$ at $A = -1.005$: those with $A < -1.005$ had a predicted treatment probability of .326, while those with $A \geq 1.005$ (but also $A < 0.046$ and $B \geq 1.047$) had a probability of .659. The remainder of the figure is read in a similar fashion. Figure 4.11 shows how the classification tree from Figure 4.10 divides the space of the predictor variables A and B into relatively homogeneous cells with respect to treatment status.

Overfitting

The in-sample performance of classification trees can be increased by increasing the number of partitions, resulting in more leaves. However, the number of cases represented by each leaf will naturally diminish as the number of partitions increases. At its logical conclusion, each leaf would contain exactly one observation, each with either a predicted probability of zero or one depending on its actual treatment status. Such a classification tree, when evaluated *within the sample*, would have perfect performance. But if one attempted to use this model to classify data from a new sample, the resulting performance would be atrocious. This problem is called **overfitting** (James et al., 2013).

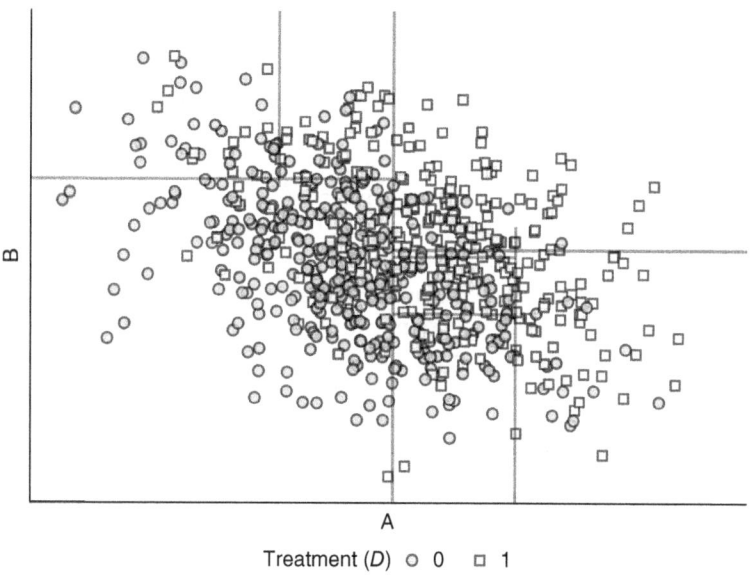

Treatment (D) ○ 0   □ 1

Figure 4.11   A classification tree partitions the data into areas of maximum class similarity

Overfitting occurs when a statistical model is allowed to learn too much from a particular sample, increasing its in-sample classification performance at the cost of out-of-sample performance (Yarkoni & Westfall, 2017). In effect, the classifier learns to discriminate on the basis of the random noise in the sample data, which does not represent real features of reality that could reliably be observed in future samples. Overfitting is a general problem in statistics and a particular bane for machine learning algorithms, where the role of a human decision-maker (hopefully) possessing theoretical insight into the phenomenon being modelled is diminished or absent. Machine learning algorithms have considerably more freedom to learn from data as compared with classical statistical models. As a result, they are particularly vulnerable. They trade the distinct possibility of overfitting for protection against the analyst's selection of a misspecified model for the data.

A word on vocabulary: because statistical models and machine learning techniques were developed in distinct intellectual traditions, they sometimes use differing terminology to describe the same underlying process. One may *fit* a statistical model to data and use it to *estimate* parameters; whereas one *trains* a machine learning algorithm, which *learns* parameter values from the data (James et al., 2013).

Boosting and **bagging** are techniques that reduce the likelihood of overfitting while at the same time increasing the performance of classification trees (Friedman, 2002). Some boosted models, such as R's gbm package, implement both of these techniques simultaneously. Both boosting and bagging are ensemble methods in which the predicted probabilities are generated by averaging over the predictions of a large set of simple classifiers. This tends to perform better than a single complex classification model at both in-sample and out-of-sample prediction when the set of classifiers is sufficiently large. The speed of modern computers makes it possible to implement ensemble techniques within a reasonable amount of time.

- Bagging methods present each of n classifiers (or 'learners') with a bootstrapped sample of the original data (Breiman, 1996). Bootstrap samples are random samples with replacement drawn from the original sample. As a result, no two classifiers see exactly the same random noise component, while the structural 'signal' in the data is constant, which reduces the tendency towards overfitting. The final predictions are based on combining the output of all the classifiers.
- Boosting methods reweight the data presented to each successive classifier, increasing the weight assigned to observations that were incorrectly classified on the basis of all the previous classifiers (Schapire & Freund, 2012). These weights raise the importance of these misclassified cases with respect to the classifier's loss function, increasing their priority. This serves to increase the performance of the ensemble. Like bagging, the final predictions are based on output from all the classifiers. However, each classifier in the sequence does not typically receive equal weight in the final predictions. A penalisation factor is applied such that later classifiers have less influence than earlier classifiers in the sequence. This factor is called the *learning rate*, *step-size reduction* or *shrinkage*, and its purpose is to reduce the tendency towards overfitting.

The analyst controls a set of hyperparameters that control the boosting process. Two of these are n, the number of classifiers to include in the ensemble, and the *shrinkage*. Others include what proportion (if any) of the sample to reserve for validation purposes, how large each tree in the ensemble is allowed to be, the minimum number of cases per leaf in the classification trees, the type of loss function (e.g. Bernoulli, AdaBoost), and the maximum number of interactions that should be considered. To use bagging, the analyst must specify the proportion of data to be resampled for each tree. And obviously, the most important hyperparameter of all is the set of variables to be included in the model.

After training a boosted classification tree model on the data, the model is then used to estimate the predicted probability of treatment for each case. These predicted probabilities are the propensity scores. Though the process for generating the scores is based on drastically different mathematics, from a pragmatic perspective, the process of obtaining the propensity scores is not so different. The biggest difference is that the analyst must tune hyperparameters rather than specify a model.

I used the gbm() function from R's gbm ('generalised boosted regression modelling') package to estimate propensity scores via boosted classification trees with an ensemble size of 5000. A cross-tabulation of the actual values of the treatment variable D versus the boosted model's predictions is displayed in Table 4.4.

Table 4.4 Boosted model-derived propensity score-predicted treatment status versus actual treatment status

	Actual $D = 0$	Actual $D = 1$
Predicted $D = 0$	312	76
Predicted $D = 1$	102	210

The model hit rate (correct classification rate, 74.6%) is, in this case, quite similar to the one obtained via logistic regression. This is because I created treatment status using a logistic regression model in the simulation – so the logistic model works quite well.

Figures 4.12 and 4.13 are produced automatically by the gbm() function. Figure 4.12 shows how the boosted model's performance varies with the ensemble size both in- and out-of-sample. Classification performance is depicted on the y-axis, with lower values of Bernoulli deviance indicating better performance. The in-sample performance continues to increase as the ensemble size grows towards 5000. However, the out-of-sample performance reaches its maximum at an ensemble size of 2427 and then worsens as additional classifiers are added. This is due to overfitting. The propensity scores should therefore be calculated from an ensemble of size 2427.

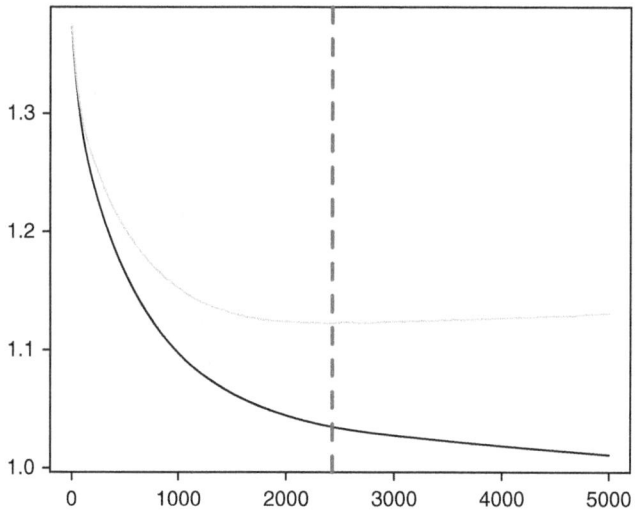

Figure 4.12 Boosted model performance in-sample (bottom line) and out-of-sample (top line) by ensemble size. The dashed vertical reference line indicates maximum out-of-sample classification performance

It is often instructive to examine how each variable in **X** is related to the model's classifications. Table 4.5 displays the relative contribution of each variable on a percentage basis. The values show that variable A makes about twice the contribution of B to the model results. These values have no absolute interpretation.

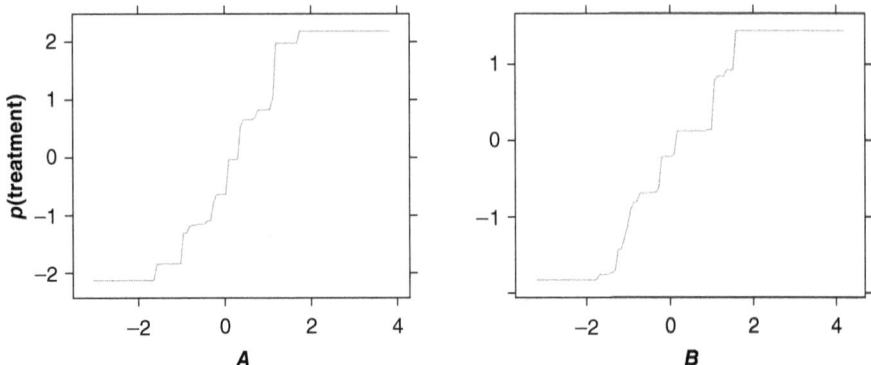

Figure 4.13 Boosted model-estimated relationships between model variables and the (logit) probability of receiving treatment

Table 4.5 The relative contribution of variables A and B to the boosted model's classifications

Variable	Relative influence
A	66.736
B	33.264

Figure 4.13 illustrates the estimated functional form of each predictor variable's relationship with the linearised (logit) probability of treatment. The jagged charac- teristic of the lines results from the discrete nature of the predicted values obtained from a classification tree model; the discontinuous jumps in the line result from the crossing of partition boundaries. Each plot describes the marginal relationship between each variable and the response. Both of these plots reveal approximately linear relationships between the variables and the logit probability of treatment. The flat areas at very high and low values of A and B result from the scarcity of extreme values in the data.

Examining these functional form plots is quite useful for gaining insight into how the boosted model arrives at predicted class membership probabilities from the data. Relationships that make little theoretical sense or possess excessively complex form can be an indication of overfitting. Note that gbm() users must pay careful attention to the units on the y-axis, because the figures can be generated with extremely trun- cated scales for variables with low relative importance. In Figure 4.13, the span of y-axis values is about twice as large for variable A compared to B, which is reflective of A's greater importance in the model's classifications.

The density plot in Figure 4.14 displays the density of propensity score by treat- ment status (D). The plot looks quite similar to the same plot produced from the logistic propensity scores (Figure 4.5) and shows that there exists sufficient common support to justify continuing with PSA.

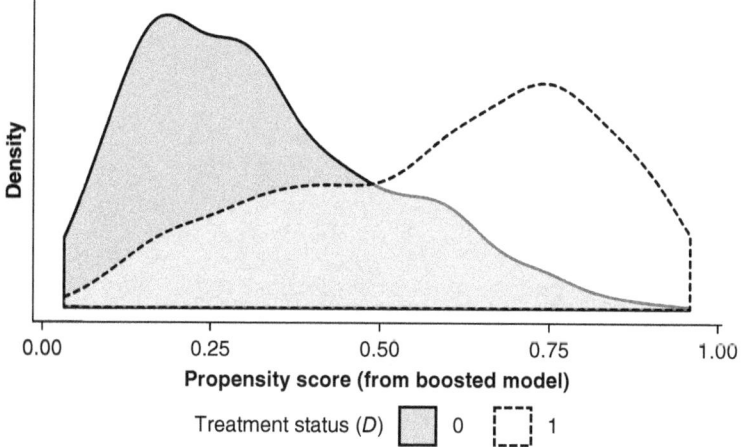

Figure 4.14 Boosted classification tree-derived propensity score density by treatment status from the analysis data

Comparing the two sets of propensity scores

Figure 4.15 contains a scatter plot of the logistic regression propensity scores against the boosted model propensity scores. The correlation between these is quite high, $r = .962$. The horizontal clustering of the boosted propensity scores is a direct result of multiple cases being assigned to the same leaf of the classification tree model and therefore having identical propensity scores. In contrast, the logistic regression scores are generated by a continuous mathematical model and exhibit no such clustering.

The question of which scores are better has an entirely pragmatic answer. The best set of propensity scores is the one that produces the optimal balancing on the **X** variables. This overrides all other concerns. It is typical for analysts to iteratively estimate propensity scores and check covariate balance, trying out different methods, models or hyperparameters, until the optimum balancing is achieved. This is permissible because most propensity score methods (and software implementations thereof) estimate the actual treatment effects as a separate step which is not performed until sufficient data balancing has been achieved. It is easy for analysts to avoid the temptation to consciously or unconsciously 'p-hack' (or select on desirability) the results regarding the treatment effect estimates across these repeated attempts. The final results are blinded by default. In fact, some software implementations (e.g. R's twang package) automate some aspects of this process.

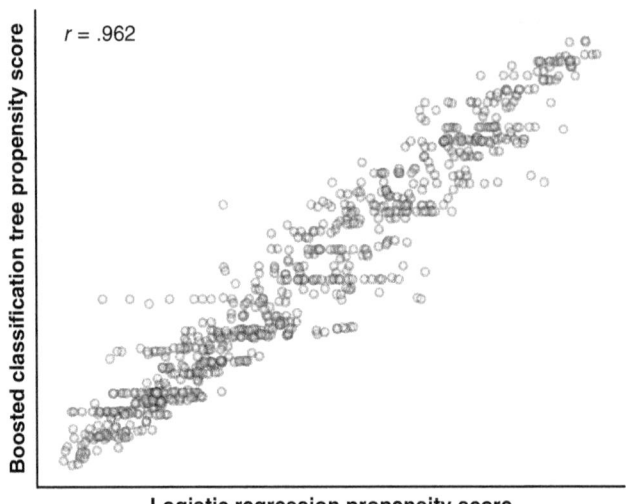

Figure 4.15 Scatter plot of logistic propensity scores versus boosted propensity scores

Assumptions of propensity score methods

Ignorability

Ignorability is the most critical assumption, and probably the hardest to fulfil. Conditional ignorability exists when the potential outcomes are independent of treatment status after conditioning on confounding variables (the **X** variables) or a summary of those variables (the propensity scores). Equivalently expressed in DAG language, the **X** variables must block all back-door paths between D and Y, rendering the potential outcomes Y_0 and Y_1 independent of D. To satisfy this condition, analysts must (a) identify all of these back-door paths, (b) precisely measure the variables necessary to close all of these back-door paths and (c) verify that the sample is sufficiently well balanced on these variables after conditioning on the propensity scores (Morgan & Winship, 2014).

Stable unit treatment value assumption

The stable unit treatment value assumption, which is often abbreviated SUTVA, states that each individual's potential outcomes are independent of other individuals' treatment assignment (Guo & Fraser, 2015). This assumption is sometimes called the 'no macro effect' or 'no interference' assumption, and it is analogous to assumption of 'no bleed-over' between conditions in randomised experiments. This condition ensures that the potential outcomes are well defined.

As an example of a SUTVA violation, imagine a study designed to evaluate the causal effect of laptop use in class on academic performances. Some students choose to use them, others do not. When a student uses a laptop, other students in the room can see the screen and could be distracted if that student is watching videos, browsing the web or engaging in other potentially distracting behaviours. This student's treatment assignment (D) can affect other students' potential outcomes. Other examples might include the following. Large-scale intervention to increase the high school graduation rate could reduce the wage benefit of high school completion through supply-and-demand and signalling effects. Second-hand smoke affects the health of nearby people regardless of whether they smoke.

The smoking example also illustrates the aspect of SUTVA regarding single versions of each treatment. In reality, smoking is not a dichotomous categorical variable. Smokers vary widely in the type and number of cigarettes they consumed, the number years of exposure they have accrued and how much second-hand smoke they expose themselves to by virtue of their behaviours. This means that $D = 1$ is not well defined as unique treatment, this creating ambiguity and variance in the potential outcomes represented by Y_1.

Positivity

The positivity assumption means that all subjects are 'at risk' of experiencing both levels of treatment. Fulfilment of this assumption implies the presence of common support across the range of the propensity scores.

$$0 < p(D = 1) < 1 \tag{4.18}$$

Violation or even near-violation of this assumption can be problematic. This assumption can be threatened due to excessive dissimilarity between groups or simply due to data sparseness. For example, propensity score methods cannot be used to estimate the causal effects of treatments (e.g. biological sex or race) that are fundamentally non-manipulable or fixed regardless of covariate values. Data sparseness can cause problems with this assumption as well, as this can lead to regions entirely devoid of treatment or control cases.

Further Reading

Little, R., & Rubin, D. B. (2010). Causal effects in clinical and epidemiological studies via potential outcomes: Concepts and analytical approaches. *Annual Review of Public Health, 21*, 121–145. https://doi.org/10.1146/annurev.publhealth.21.1.121

This article provides a readable and relatively non-technical introduction to Rubin's causal model, situating it within a discussion of randomization-based inference.

Rosenbaum, P. R., & Rubin, D. B. (1983). The central role of the propensity score in observational studies for causal effects. *Biometrika, 70*(1), 41–55.
This is one of the seminal articles on propensity scores. It proposes and proves five theorems which show that balancing samples on the propensity scores is sufficient for the unbiased estimation of causal effects.

Rubin, D. B. (1997). Estimating causal effects from large data sets using propensity scores. *Annals of Internal Medicine, 127*, 757–763. https://doi.org/10.7326/0003-4819-127-8_Part_2-199710151-00064
This article provides another non-technical introduction to propensity scores, describing how they can be used in the context of large observational datasets to draw more reliable conclusions.

Rubin, D. B. (2005). Causal inference using potential outcomes: Design, modeling, decisions. *Journal of the American Statistical Association, 100*(469), 322–331. https://doi.org/10.1198/016214504000001880
This article is a transcribed version of an address Rubin delivered at the 2004 Joint Statistical Meeting. In it, he situates his perspective in Fisher's prior work on experimental design and focuses on the key role played by the assignment mechanism in causal inference.

West, S. G., & Thoemmes, F. (2010). Campbell's and Rubin's perspectives on causal inference. *Psychological Methods, 15*(1), 18–37. https://doi.org/10.1037/a0015917
This article considers the similarities and differences between Campbell's threats to validity framework and Rubin's counterfactual framework.

5

PROPENSITY SCORE ANALYSIS

Chapter Overview

The concept of propensity scores, their connection to RCM and methods for estimating them were introduced in Chapter 4. This chapter describes three methods for using propensity scores for constructing inferences: subclassification, weighting and matching. These conditioning methods were described in general terms in Chapter 2. Techniques that involve the use of propensity scores in estimating causal effects are collectively known as propensity score analysis (PSA).

Simulated data set

The examples in this chapter will be based on data simulated from the DAG shown in Figure 5.1. This is the same data set that was used in Chapter 4 to illustrate propensity score estimation, except that the propensity scores derived from the logistic and boosted models have been appended to it. In this data, variables A and B are confounders, D is a binary treatment (exposure) variable and Y is a continuous response variable. Most real applications of PSA will involve a larger set of variables, and all the techniques presented in this chapter can be extended to much larger variable sets without loss of generality. Limiting these examples to two variables makes it possible to represent the various PSA methods graphically, which I have found helpful in developing an intuitive understanding of how they work. The data set can be downloaded from this book's companion website.

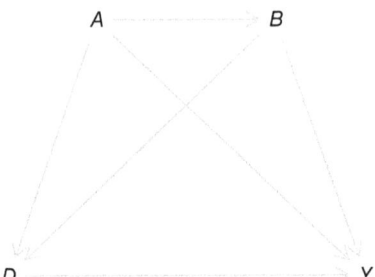

Figure 5.1   Example of a DAG for the propensity score analysis examples

Note. DAG = directed acyclic graph.

Although the example analyses in this chapter will be performed in R, the Stata statistical software also has excellent capability for PSA. I particularly like psmatch2 for matching (Leuven & Sianesi, 2018).

The true values for the average treatment effect (ATE), average treatment effect for the treated (ATT) and average treatment effect for the untreated (ATU) in the synthetic data set are all 0.3.

Descriptive statistics and biased treatment effect estimate

The first step in any analysis is to examine the data. Table 5.1 contains descriptive statistics by treatment group, Figure 5.2 displays density plots of the confounding variables A and B by treatment status and Figure 5.3 displays a scatter plot and correlation coefficient for A versus B. Since there is an edge connecting A and B in the DAG, these variables are causally connected and must therefore be correlated. The scatterplot displayed in Figure 5.3 confirms that they are.

Table 5.1 Descriptive statistics by treatment group for synthetic data

Variable	D	Valid n	Mean	SD	Minimum	Maximum
A	Treatment	312	0.41	0.97	−2.13	3.81
	Control	388	−0.36	0.92	−3.01	2.50
B	Treatment	312	0.16	1.06	−3.17	4.18
	Control	388	−0.13	1.00	−2.43	2.67

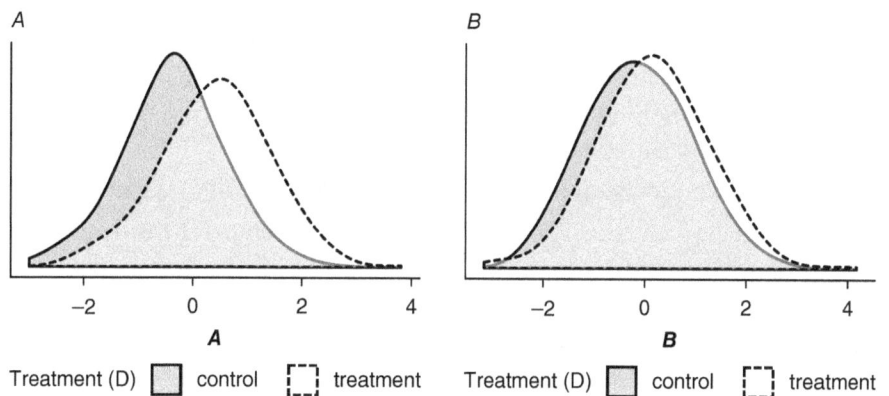

Treatment (D) ▨ control ⬚ treatment   Treatment (D) ▨ control ⬚ treatment

Figure 5.2 Frequency distributions of variables A and B by treatment condition (D)

Obtaining a biased estimate of the treatment effect

As described in Chapter 4, RCM defines causal effects in terms of a mixture of observable quantities and unobservable counterfactuals. Under the condition of ignorability, values calculable from observed data can replace the missing counterfactuals. This allows researchers to generate unbiased estimates of these fundamentally unobservable causal effects.

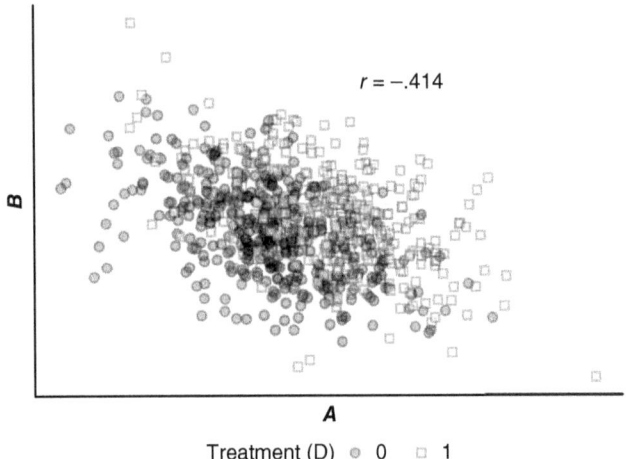

Figure 5.3   Scatter plot and correlation coefficient of A versus B

Ignorability means that the potential outcomes are independent of treatment assignment. Conditional ignorability means that ignorability is achieved after conditioning on an appropriate set of background variables. If neither ignorability nor conditional ignorability is achieved, biased causal estimates will result (Guo & Fraser, 2015). From the DAG (Figure 5.1), it is clear that ignorability does not exist due to the existence of confounders A and B. It is also apparent that conditional ignorability would require conditioning on both of these variables. Conditioning on either of them alone is insufficient to block all the confounding paths.

The naive estimator of the treatment effect is simply an unadjusted, unconditional comparison of $Y_1|D = 1$ versus $Y_0|D = 0$, the observed outcomes for the treatment and control groups. This quantity can be calculated via an independent samples t-test or a linear regression model. Both will, of course, yield identical results. When no conditioning variables are considered, the only treatment effect that can be estimated is the ATE. The ATT and the ATU cannot be estimated without incorporating background variables into the analysis, because these are moderated treatment effects.

Biased model results

Without conditioning on variables A and B, the estimated treatment effect will be biased. Here I will show two methods for estimating this biased effect, which will be contrasted with the unbiased estimates that are obtained using propensity score methods. Since the exposure or treatment variable D is dichotomous and the outcome Y is continuous, a simple t-test can be used to estimate the treatment effect as well as its confidence interval and p-value. The output from R's t.test() function is as follows.

```
##
## Two Sample t-test
##
## data: Y by D
## t = -5.4127, df = 698, p-value = 8.545e-08
## alternative hypothesis: true difference in means is not equal to 0
## 95 percent confidence interval:
##  -0.7688856 -0.3595592
## sample estimates:
## mean in group 0 mean in group 1
##       -0.1312448       0.4329776
```

The estimated ATE according to this naive approach is calculated as the difference in the observed group means, which are reported in the last line of the output: $0.433 - (-0.131) = 0.564$. The hypothesis test suggests that the null hypothesis of zero treatment effect should be rejected, $t(698) = -5.413$, $p < .001$. The 95% CI around the estimated treatment effect is [0.360, 0.769]. The true value of 0.3 is not included in this interval.

The same estimate can be produced by fitting a simple regression of Y on D to the data. I used R's lm() function to do this. Here are the results:

```
##
## Call:
## lm(formula = Y ~ D, data = data)
##
## Residuals:
##     Min      1Q  Median      3Q     Max
## -3.9440 -0.8583  0.0252  0.8702  4.2770
##
## Coefficients:
##               Estimate Std. Error t value Pr(>|t|)
## (Intercept) -0.13124    0.06959  -1.886   0.0597 .
## D            0.56422    0.10424   5.413 8.54e-08 ***
## ---
## Signif. codes:  0 '***' 0.001 '**' 0.01 '*' 0.05 '.' 0.1 ' ' 1
##
## Residual standard error: 1.371 on 698 degrees of freedom
## Multiple R-squared: 0.04028, Adjusted R-squared: 0.03891
## F-statistic: 29.3 on 1 and 698 DF, p-value: 8.545e-08
```

The slope coefficient for D is the estimated ATE. Note that its test statistic and p-value match the values reported in the t-test output. Also note the equivalence in the residual standard error degrees of freedom to the t-test statistic's degrees of freedom.

Propensity score matching

If the sources of confounding can be identified and the variables necessary to block it can be measured with high precision, a conditioning-based strategy can be employed to estimate the causal effect. If there are multiple variables in the minimum adjustment set (see Chapter 3 for discussion), as is usually the case, then PSA can be used to solve the 'curse of dimensionality' that affects the non-parametric conditioning methods such as matching, subclassification and weighting. The propensity scores distil the most relevant information in this potentially large set of conditioning variables down to a single variable, which makes it possible to implement these non-parametric techniques in realistic situations. The use of any of these methods can approximate the same 'equivalence in expectation' condition with respect to these variables that would have occurred under a randomised design so long as the assumptions of PSA are met (Guo & Fraser, 2015).

Propensity score matching is perhaps the most popular method for conditioning on the propensity score, and certainly the most intuitive. The goal of matching is to create a virtual counterfactual observation for each target case. For example, a target treatment case is paired with a control case that has a similar value of the propensity score. Equivalence on the propensity score implies that this matched case is equivalent to the target case on $\mathbf{X}$ in expectation. The observed outcome for this matched case provides an unbiased estimate of what the target case's outcome *would* have been under the control condition (or, more generally, the other level of the exposure variable). Matched cases allow for virtual peeks at this fundamentally observable counterfactual. The matched case's observed outcome is not equivalent in value to the counterfactual, but it is drawn from the same distribution as the actual counterfactual would have been (Ho et al., 2007). In other words, the matched case's observed outcome may be higher or lower than the target case's actual counterfactual, but over the long run, the average error is zero. And this means that the aggregated treatment effects, such as the ATE, ATT and ATU, can be estimated without bias by averaging over a large set of individual treatment effects computed as the difference between the target and matched cases' observed outcomes.

Matching algorithms

A tremendous number of matching algorithms are available, and the decision space is huge. The advantage is that analysts have great freedom to tinker with the process to try to achieve the best possible covariate balance. The dark side of this freedom is the possibility of selecting on the treatment effect estimates, because the estimated treatment effects may vary over matching procedures (Baser, 2006). Nearly all propensity score software tools separate the balance assessment process from the treatment effect estimating process, enabling researchers to repeat the former without knowledge of the latter. (Contrast this with linear regression, where the results are typically presented before diagnostic criteria are made available.) My advice is that, if you are operating in a confirmatory (rather than exploratory) research context, you should avoid examining the estimated treatment effect until you are completely satisfied with the quality of covariate balance. Once you have seen the estimated treatment effect, the analysis should be frozen. No further tweaks are permissible without entering the realm of exploratory research (McBee & Field, 2017). There is no going back.

In this section, I will describe some popular well-known matching algorithms and discuss their strengths and weaknesses.

One-to-one versus one-to-many matching

In one-to-one matching, each target case is matched to one and only one case from the other group. In contrast, one-to-many matching allows $k > 1$ cases to be matched to each target case, where k is specified by the analyst (Rassen et al., 2012). One-to-many matching is most useful when the data are substantially imbalanced, as often occurs when studying rare or unusual circumstances. For example, there are many more high school completers than there are dropouts. One-to-many matching in such a circumstance can allow more of the available data to be used. Figure 5.4 illustrates one-to-one versus one-to-many matching. In this example, there are three control cases available for matching to one treatment case. One-to-one matching discards two-thirds of the control cases, as A and C are unmatched, whereas one-to-many matching uses all available controls.

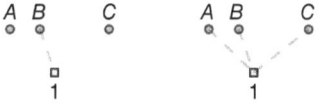

Figure 5.4 One-to-one (left panel) versus one-to-many (right panel) matching

In one-to-many matching, the multiple matched cases create a single 'virtual' case to join to the target case. In the right panel of Figure 5.4, case 1 is matched to cases A, B and C. The virtual matched case for case 1 is $(1/3) A + (1/3) B + (1/3) C$. These numbers are weights, and they are necessary because the pairings involving a shared element are not independent and contribute a smaller amount of information. These weights must be incorporated into the analysis to obtain proper inference (Stuart, 2010).

Greedy versus optimal matching

Figure 5.5 illustrates the difference between the greedy and optimal versions of one-to-one nearest-neighbour matching without replacement. The treated cases are represented by white squares, control cases by grey circles. If the researcher is interested in estimating the ATT, the greedy matching algorithm sequentially visits each treatment case and assigns to it the closest available control case (Guo & Fraser, 2015). The greedy algorithm's matches are displayed in the upper panel of the figure. The total mismatch, computed as the sum of the absolute value of the propensity score mismatch over all four matched pairs, is 0.71. Note that this number has no absolute interpretation as it depends on the sample size, but it is useful for comparing the performance of matching algorithms.

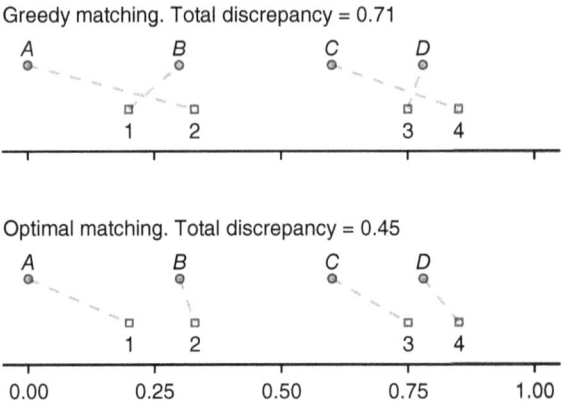

Figure 5.5 Greedy versus optimal matching. Dashed lines connect matched units. Optimal matching leads to higher aggregated similarity between the matched pairs

Optimal matching is a computationally expensive algorithm that attempts to find the set of matches that results in the smallest aggregated mismatch (Ming & Rosenbaum, 2001). Optimal matching results are displayed in the lower panel of Figure 5.5. Case 1 is matched to A even though B is closer. This allows case 2 to be matched with B, which produces much greater similarity than if case 2 were matched with A. Similarly, case 3 is matched to C even though D is closer, allowing 4 to be matched to D instead of C. The total mismatch in this case is 0.45, which is a 58% improvement in the overall similarity of the matched cases.

In addition to the generally poor performance of greedy matching algorithms, the sequential nature of the algorithm means that the order in which the cases are matched matters. Some software, such as Stata's psmatch2 routine, match the cases sequentially in the order that they appear in the data set (Leuven & Sianesi, 2018). Randomly sorting the data set prior to greedy matching is often recommended, as the specific pairings and hence the resulting treatment effect estimates may vary with sort order. This is highly undesirable behaviour from the standpoint of reproducibility. It is difficult to say that any particular estimate is *the* result of the analysis.

Caliper matching

Caliper (or radius) matching refers to a type of one-to-many matching in which all of the available cases within a specified radius are matched to the target case. The analyst does not specify a fixed value for k, and the number of matches for each target varies (Huber et al., 2015). The caliper (or radius) is the maximum allowed dissimilarity or distance between matched units, specified in propensity score units. For example, a caliper of 0.1 means that treatment and control cases must be within 0.1 units of one another on the propensity score in order to be joined together. Caliper matching is effectively a type of one-to-many matching in which the number of cases to be joined to each target case can vary. Cases that cannot be matched within the caliper are discarded. Specifying a caliper prevents highly dissimilar cases from being matched and is helpful for reducing bias in the treatment effect estimates (Austin, 2011). In other words, it confines the matching process to the region of common support, which is highly desirable. How tight should the caliper be? Narrowing the caliper results in closer overall matches at the expense of sample size. The caliper width is a classic bias–variance trade-off. Pragmatically, it is often set to the highest value that produces subjectively reasonable balancing on the propensity score and on the X variables as identified through iterative experimentation.

Kernel matching

Kernel matching is a type of one-to-many matching where variable weights are assigned to each matched case depending on their distances from the target (Apel & Sweeten, 2010). For example, in the lower panel of Figure 5.4, cases A, B and C all received weights of 1/3 even though B is the closest to case 1 and C is the furthest. The quality of the 'virtual' counterfactual observation for the target case can be improved by differentially weighting the matched cases in proportion to their similarity or distance. Cases with higher similarity to the target should receive more weight. The virtual counterfactual is then a weighted average of the matched cases.

There are numerous choices of weighting function. These functions are called kernels. Kernel matching typically uses a caliper to enforce a minimal level of similarity between matched cases. A uniform kernel assigns equal weight to all the matched cases within a caliper. Uniform kernel matching is therefore the same as caliper matching. A Gaussian kernel envisions a normal distribution around each target case.

Weights are assigned on the basis of the relative density. Figure 5.6 displays the Gaussian (normal), Epanechnikov (parabolic) and triangular kernel functions and the normalised weights derived from each. These weights have been normalised to sum to one for each target case. For example, prior to normalisation, the Gaussian weights were 0.773, 1.748 and 0.598. The sum of these weights is 1.766. Each weight was divided by this sum to yield the normalised weights of 0.248, 0.56 and 0.192. Normalisation is necessary to ensure that each virtual counterfactual, constructed as a weighted sum of the matched cases, is worth the same amount (in terms of information contributed) as an actual observation of the counterfactual.

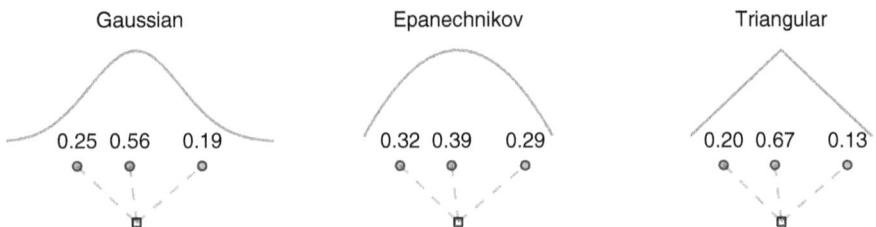

Figure 5.6 Illustration of three kernel functions. Left: Gaussian. Middle: Epanechnikov. Right: Triangular. The numbers are the normalised weights assigned to each matching case

Most kernel functions have adjustable parameters that can strongly affect the relative weights (Apel & Sweeten, 2010). For example, the Gaussian function has a standard deviation parameter that controls the 'spread' of the curve and thus how quickly the weights decay as a function of distance from the target. The choice of such parameters can be more consequential than that of the kernel function itself. Moreover, some kernel functions, such as the Gaussian, produce positive non-zero weights for all the cases unless a caliper is imposed. Without a caliper, the Gaussian kernel matches each case to a weighted combination of *all* the other cases. The other two kernel functions in Figure 5.6 do not exhibit this behaviour, naturally crossing zero at some distance from the target case.

There are many other kernel functions besides these. All serve to control the decay rate of the relative weights as a function of distance from the target. Like one-to-many matching, the weights must be incorporated into the analysis.

Matching with replacement

The default matching process is matching without replacement. Once a case has been matched to a target case, it becomes unavailable for further matching. Matching with replacement allows cases to be matched more than once. Figure 5.7 shows this process. The left panel of the figure illustrates greedy one-to-one matching without replacement.

The matching algorithm begins with treatment case 1 and matches it to control case B. Next, case 2 is matched to case C, the best remaining match in the sample.

Figure 5.7 Matching without replacement (left) versus matching with replacement (right)

The right panel shows matching with replacement. As before, treatment case 1 is matched to control case B. But case B is still available and is selected again as the closest match for 2. This can lead to superior overall balancing on the propensity score and therefore on **X** in expectation. However, since B was selected twice, the pairings involving it must receive a weight of 1/2 because they are not independent.

Matching with replacement is quite similar to one-to-many matching. Suppose that a researcher has a set of treatment cases and seeks to join them to a set of matched controls. One-to-many matching allows multiple control cases to be matched to each treatment case, whereas matching with replacement allows multiple treatment cases to be matched to each control case (Rassen et al., 2012). The distinction involves which cases may be repeated. These two methods may also be combined, allowing repetition on both the treatment and control sides.

Estimating treatment effects with matching

Matching may be used to calculate any of the three treatment effects. The ATT can be estimated on the basis of a matched sample in which a matched control case (or cases, depending the matching algorithm) is selected for each treated case. The ATU can be estimated by finding one or more treatment cases for each control case. The distinction lies with which set of cases are viewed as the observed potential outcomes and which are the virtual counterfactuals. To estimate ATT, start with treatment cases and find matching controls. To estimate ATU, start with control cases and find matching treated cases. The ATE is the weighted average of these two conditional treatment effects (Stuart, 2010).

Example analysis

I used the R package Matching to estimate the ATE using one-to-one matching on the logistic propensity scores with replacement, enforcing a radius of 0.03. The Match() function performs the matching, and the MatchBalance() function diagnoses the achieved covariate balance. Its output is given below.

```
## 
## ***** (V1) A *****
##                              Before Matching           After Matching
## mean treatment........       0.41039                   -0.026994
## mean control.........        -0.36374                  -0.061693
## std mean diff.........       79.582                    3.8388
## 
## mean raw eQQ diff.....       0.78108                   0.091767
## med  raw eQQ diff.....       0.799                     0.065
## max  raw eQQ diff.....       1.312                     0.527
## 
## mean eCDF diff........       0.21897                   0.024389
## med  eCDF diff........       0.23533                   0.012937
## max  eCDF diff........       0.3716                    0.087969
## 
## var ratio (Tr/Co).....       1.1256                    0.90805
## T-test p-value........     < 2.22e-16                  0.36757
## KS Bootstrap p-value..     < 2.22e-16                  0.002
## KS Naive p-value......     < 2.22e-16                  0.0050481
## KS Statistic..........       0.3716                    0.087969
## 
## 
## ***** (V2) B *****
##                              Before Matching           After Matching
## mean treatment........       0.16099                   -0.050096
## mean control.........        -0.13449                  0.00049011
## std mean diff.........       27.759                    -4.743
## 
## mean raw eQQ diff.....       0.31859                   0.091018
## med  raw eQQ diff.....       0.316                     0.058
## max  raw eQQ diff.....       1.516                     1.516
## 
## mean eCDF diff........       0.081726                  0.017526
## med  eCDF diff........       0.08862                   0.01423
## max  eCDF diff........       0.14416                   0.063389
## 
## var ratio (Tr/Co).....       1.1246                    1.0537
## T-test p-value........     0.00019709                  0.37892
## KS Bootstrap p-value..     < 2.22e-16                  0.069
## KS Naive p-value......     0.0015105                   0.089544
## KS Statistic..........       0.14416                   0.063389
## 
## 
## Before Matching Minimum p.value: < 2.22e-16
## Variable Name(s): A B  Number(s): 1 2
## 
```

```
## After Matching Minimum p.value: 0.002
## Variable Name(s): A  Number(s): 1
## [1] "2"
```

This output reveals that the matching process removed much of the imbalance on variables A and B. For variable A, the significant Kolmogorov-Smirnov (KS) bootstrap p-value after matching paired with its non-significant t-test indicates that the mean of variable A does not significantly vary across groups after matching, but that the overall distribution of A is not identical across matched groups. The balance on variable B is better; both its t-test and KS tests reveal no significant residual imbalance in mean or distribution after matching. The treatment effect estimate is produced by running summary() on the object containing the Match() function's results. The output is given below:

```
##
## Estimate...  0.38929
## AI SE......  0.12897
## T-stat.....  3.0185
## p.val......  0.0025402
##
## Original number of observations.............  700
## Original number of treated obs..............  312
## Matched number of observations..............  632
## Matched number of observations  (unweighted).  773
##
## Caliper (SDs)..............................0.03
## Number of obs dropped by 'exact' or 'caliper'  68
```

The estimated ATE is 0.389 with 95% CI [0.137, 0.642]. The true value for the ATE of 0.3 is contained in this interval.

Stratifying on the propensity score

As described in Chapter 2, subclassification (*stratification*) is a simple method for conditioning on a variable. The sample is divided into subsets (*strata*) with similar values of the propensity score. The treatment effect is calculated separately within each stratum and then aggregated into an overall or marginal value: the ATE.

Once the propensity scores have been estimated using the logistic regression approach, the boosted classification tree approach or some other method, the first decision in the analysis is to choose the number of strata. There are many methods for this, including equal intervals and equal percentiles (which ensure that each stratum contains approximately the same sample size), among others. One must ensure that each stratum contains a reasonable number of treatment and control cases. There is an unavoidable bias–variance trade-off in choosing the number of strata, as increasing the number reduces bias but will quickly lead to sparsely populated cells, increasing variance. Cochran (1968) showed that five strata are generally sufficient for removing 90% or more of the bias. Beyond that, the 'right' choice in stratum creation is the minimum number required to minimise residual covariate imbalance. It is perfectly acceptable to assess the consequences of several different sets of strata before moving forward with the analysis, or to perform multiple version of the analysis using different numbers of strata as a sensitivity check.

Based on the logistic regression propensity scores (variable pscores in the data set), I created $k = 5$ strata with boundaries at the 20th, 40th, 60th and 80th percentiles using the quantile() and cut() functions in R, resulting in $n = 140$ per stratum. Table 5.2 displays the number of treated and untreated cases in each. Figure 5.8 shows the boundaries of the strata. It is obvious from both that the proportion of untreated cases is higher in the low-propensity score strata and higher in the high-propensity score strata. This will always be the case when the propensity score model has reasonable classification accuracy.

Table 5.2 Treatment and control cases by stratum

Stratum	D	n
1	0	119
	1	21
2	0	104
	1	36
3	0	88
	1	52
4	0	58
	1	82
5	0	19
	1	121

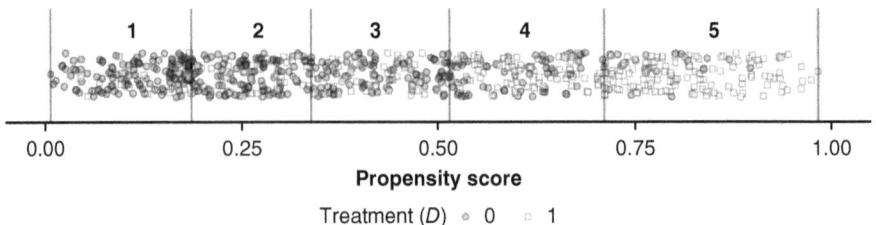

Figure 5.8 Stratum boundaries

The R package PSAGraphics is an excellent tool for visualising balance and calculating treatment effects using subclassification. Within each stratum, the propensity scores for the two groups are approximately the same by virtue of their restricted range. Conditioning on the propensity scores in this manner implies that all of the **X** variables in the propensity score model should be balanced in expectation across groups. If true, this means that a straightforward comparison of observed treatment and non-treatment means *within strata* is an unbiased estimator of the causal effect of treatment. It is crucial to assess whether this balance was actually achieved before proceeding to estimate and interpret the effects.

Figure 5.9 was generated by the box.psa() function in this package and visually displays the covariate balance within strata. A box-and-whisker plot of the data for the treated and untreated groups provides a visual depiction and summary of the distributions with the (jittered) actual data points overplotted. Obviously, these should be as similar as possible; any remaining discrepancy can bias the treatment effect estimate. The numbers beneath each box plot are the number of cases, and reference lines connect the treatment and untreated means. A similar function, cat.psa(), creates similar plots for categorical variables.

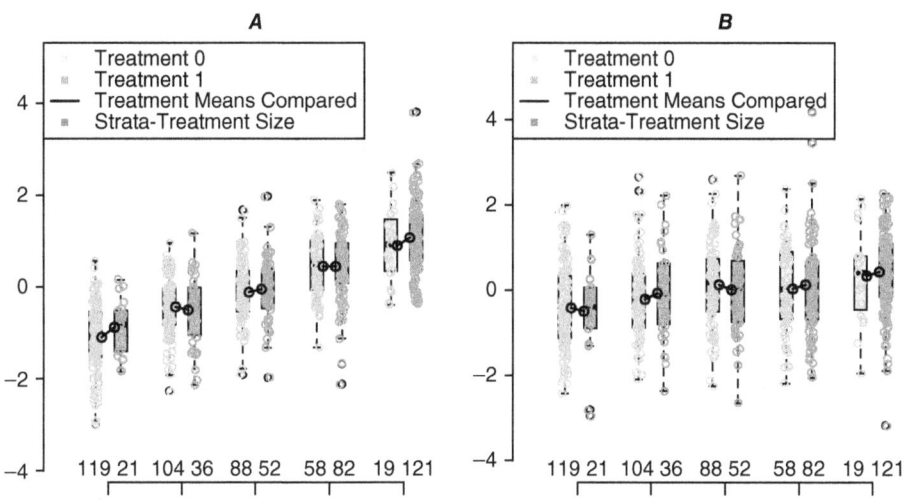

Figure 5.9 Covariate balancing for variables *A* (left panel) and *B* (right panel) within strata

The means for each group within strata for *A* and *B* are shown in Table 5.3. This table along with Figure 5.9 show that the subclassification procedure produced reasonably good covariate balance. (Table 5.3 shows the baseline mean differences as a basis for comparison.) In a typical application of propensity score subclassification, the analyst would likely evaluate this balance information for several different configurations of strata and propensity score estimation techniques in search of the best balance that can be achieved.

Table 5.3 Balance table: treatment and control means for variables A and B by stratum

Stratum	D	Mean (A)	Mean (B)
1	0	−1.09	−0.42
	1	−0.88	−0.49
2	0	−0.43	−0.21
	1	−0.50	−0.07
3	0	−0.11	0.13
	1	−0.05	0.01
4	0	0.45	0.03
	1	0.45	0.13
5	0	0.90	0.34
	1	1.08	0.43

The `circ.psa()` function from PSAGraphics calculates the treatment effect and produces an exceptionally informative graphical summary of the results, which is displayed in Figure 5.9. The outcome under the untreated (control) condition (Y_0) is on the x-axis, the outcome under treatment (Y_1) on the y-axis. The horizontal reference line stretching from the lower-left to the upper-right corner of the plot is the line of zero treatment effect, in which the outcomes under treatment and control have the same value. The vertical and horizontal dotted lines display the grand-mean values of Y_0 and Y_1

Each stratum is represented by a circle whose centre is located at its pairing of ($E[Y|D = 0]$, $E[Y|D = 1]$) values. If the propensity score assumptions are met, these values estimate the expected values of the potential outcomes ($E[Y_0]$, $E[Y_1]$). The difference in these values is the estimated treatment effect within the stratum. The circles will be located above the line of zero effect when the treatment effect is positive (implying $Y_1 > Y_0$) and below it otherwise. In this case, positive treatment effects are observed for all of the strata. The circles are sized proportionally to the stratum sample size; they are identical in this figure because all these strata contain $n = 140$.

The diagonal, heavy dotted line displays the ATE, which is calculated as an average of all the stratum-specific treatment effects weighted by their sample sizes. Its offset from the line of zero effect displays the magnitude of the treatment effect. The thick diagonal, line with end caps, displays the boundaries of the 95% CI computed around the ATE estimate. When these boundaries exclude the line of zero effect, one can conclude that a two-tailed hypothesis test would reject the null hypothesis of no treatment effect at the $\alpha = .05$ level.

In addition to the image in Figure 5.10, the `circ.psa()` function produces additional output, which is as follows. The `$summary.strat` component gives the sample sizes and means under treatment and control within strata, which are the coordinates of each circle in the figure. The weighted means under treatment and control across strata are $wtd.Mn.0 and $wtd.Mn.1. The estimated ATE, along with its standard error, approximate *t*-test statistic and degrees of freedom, and 95% CI, are also reported. In this case, the estimated ATE is 0.409 with 95% CI [0.155, 0.662]. The true value for the ATE of 0.3 is contained in this interval.

```
## $summary.strata
##    n.0 n.1     means.0    means.1
## 1 119   21 -0.37915126 0.3366667
## 2 104   36 -0.13869231 0.1618333
## 3  88   52 -0.10111364 0.2420962
## 4  58   82  0.27420690 0.4275244
## 5  19  121  0.08494737 0.6160909
##
## $wtd.Mn.0
## [1] -0.05196059
##
## $wtd.Mn.1
## [1] 0.3568423
##
## $ATE
## [1] 0.4088029
##
## $se.wtd
## [1] 0.1292062
##
## $approx.t
## [1] 3.163957
##
## $df
## [1] 690
##
## $CI.95
## [1] 0.1551184 0.6624874
```

Though this discussion has focused on using subclassification to estimate the ATE, it can also be used to estimate other treatment effects such as the ATT and the ATU. These are estimated by differentially weighting the contribution of each stratum-specific treatment effect. The stratum-specific effects can be weighted based on the number of *treatment* cases they contain to estimate ATT and on the number of *control* cases to estimate the ATU. Weighting the stratum treatment effects in proportion to the total sample size each one contains estimates the ATE (Guo & Fraser, 2015).

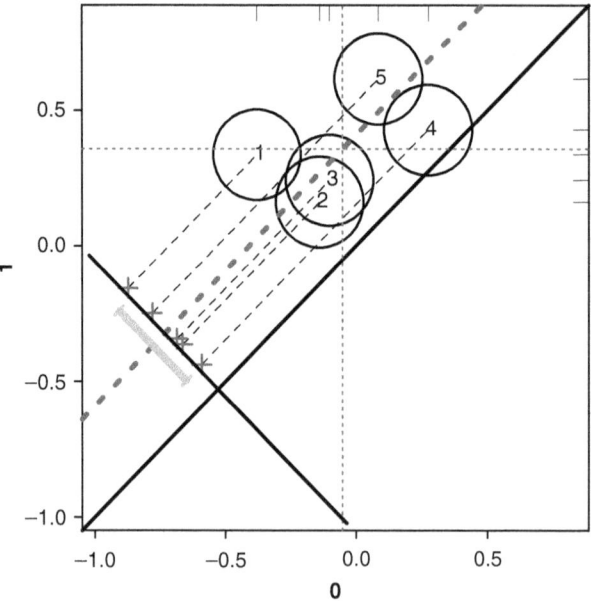

Figure 5.10 Stratified propensity score analysis summary

Weighting with the propensity score

Propensity scores may also be used as weights in order to estimate causal effects. This approach offers some compelling advantages over subclassification. First, the number of decisions that must be made by the researcher is quite minimal. Second, weighting can be used to estimate the conditional causal effects, ATT and ATU, as well as the ATE. These features make propensity score weighting a compelling technique for researchers who want to reduce their researcher degrees of freedom (Wicherts et al., 2016) to a minimum. This small decision space makes propensity score weighting particularly advantageous for preregistered studies (Nosek et al., 2018). The researcher needs to only prespecify the **X** variables (ideally informed by a DAG), the method for estimating the propensity scores, the treatment effect(s) to be estimated, the software to be used and the criteria for satisfactory common support and covariate balance.

The intuition underlying the weighting approach is illustrated in Figure 5.11. The underlying cause of the bias in the estimated treatment effects is the non-equivalent distributions of the propensity scores for the treated and untreated groups. The top panel of Figure 5.11 displays this imbalance.

If the researcher wants to estimate the ATT, the distribution of control cases must be reweighted in order to make it resemble the distribution of treated cases on the propensity scores. This creates a 'virtual population' of cases in which the distribution of propensity scores has been equalised between the treatment and control groups, allowing the unbiased estimation of treatment effects. The middle panel of

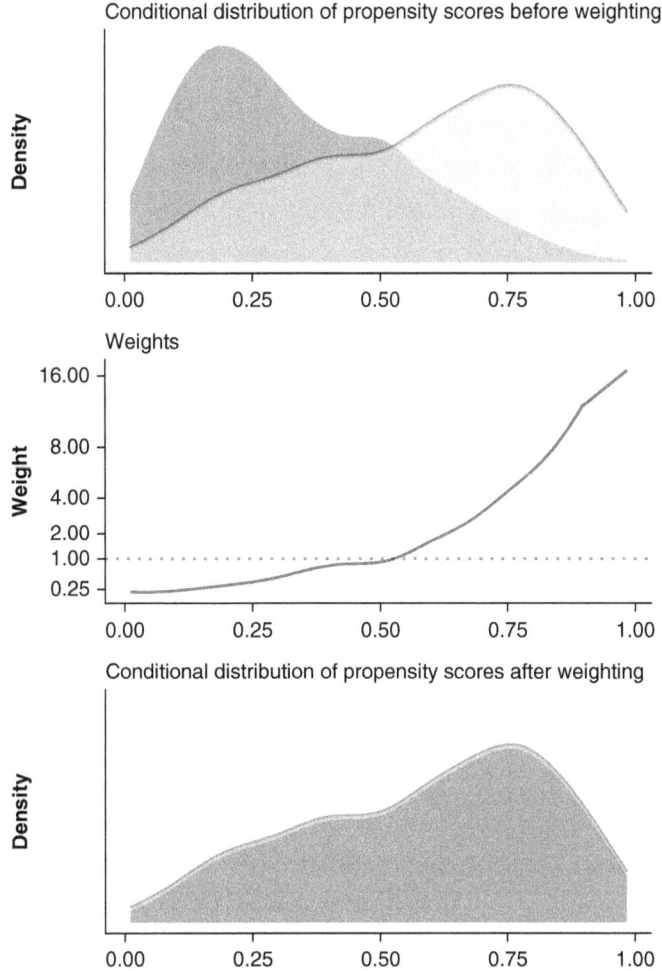

Figure 5.11 Illustration of using weights to estimate the ATT. Top panel: The unweighted distribution of propensity scores for the untreated (left) and treated (right; target distribution). Middle panel: The weights applied to the untreated cases. Note the non-linear scale for the y-axis. Bottom panel: The weighted distribution of propensity scores. After weighting, the distribution of propensity scores for the untreated resembles the distribution for the treated

Figure 5.11 shows these weights, which are < 1 where the density of control cases is higher than treated cases and > 1 in regions where the density of control cases is lower. The bottom panel shows the result of applying the weights. The distribution of control cases is now equivalent to the distribution of treated cases on the propensity scores. Variants of this strategy, which will be explained later in this chapter, can be used to estimate the ATE or the ATU.

Figure 5.12 shows the conditional distributions of variables A and B by treatment condition before and after weighting. In this case, the weighting procedure has pushed the distributions of A and B in the control condition to resemble their distributions in the treatment condition. A treatment effect computed using this set of weights would estimate the ATT, because the distribution of all the **X** variables now resembles that of the treatment group.

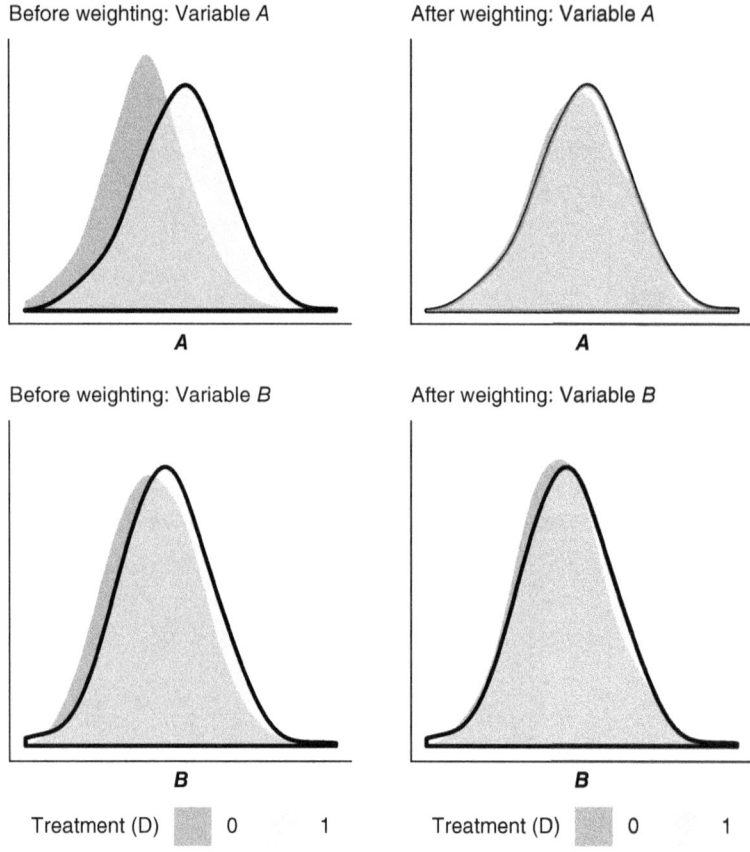

Figure 5.12 Applying the weights to the sample balances *A* and *B* in expectation across treatment conditions

It is clear from Figure 5.12 that the weights removed a large proportion of the imbalance, but that some remains, particularly on variable *B*. In theory, weighting the sample to equalise the distribution of propensity scores also perfectly balances the sample in expectation on all of the **X** variables. In practice, the propensity scores are themselves estimated with uncertainty, and the propensity score model itself may suffer various imperfections such as misspecification bias or measurement error on the covariates. As a result, the weighting procedure will not achieve exact balance **X** in practice. (Note that this is common to all propensity score methods and is not specific to weighting procedures.) The question is whether the covariate balance in the virtual population is good enough to justify moving forward with the estimation and interpretation of causal effects.

Some residual imbalance on the covariates will always persist after conditioning on the propensity score. This residual imbalance is directly proportional to the potential bias in the estimated treatment effects generated from the analysis. Researchers must therefore carefully assess the achieved balance on all the **X** variables to determine whether the equivalence is good enough to justify proceeding with the analysis. PSA software routines provide numeric or graphical tools for assessing balance.

From propensity scores to weights

The weights applied in weighted PSA are calculated from the propensity scores. These weights are called inverse probability of treatment weights (IPTW). Different sets of weights are used to estimate the various treatment effects, but the weights used for estimating ATE make very clear just how apt this term is. The equation for calculating the ATE weights is

$$W_{ATE} = \frac{D}{p(D=1|\mathbf{X})} + \frac{1-D}{1-p(D=1|\mathbf{X})} \tag{5.1}$$

where D is a binary 0/1 treatment indicator and $p(D = 1|\mathbf{X})$ is the propensity score – the conditional probability of being treated given **X**. The equation is written such that one of the terms drops out depending on the value of D.

$$W_{ATE} = \begin{cases} 1/p(D=1|\mathbf{X}) & (D=1) \\ 1/[1-p(D=1|\mathbf{X})] & (D=0) \end{cases} \tag{5.2}$$

For estimating the ATE, the weights applied to the treatment group are the reciprocal of their propensity scores, while the weights applied to the untreated group are the reciprocal of one minus their propensity scores. Since the propensity score is conditional probability of receiving treatment, and it is assumed that all cases are either treated or untreated, one minus that quantity is the conditional probability of being untreated. So the weights for both groups are simply the reciprocal of the conditional probabilities of being in the group they are in.

The justification for this weighting scheme is perhaps easier to grasp by example. A treated case with a high propensity score of $p = .9$ would receive a weight of $1/.9 = 1.111$. Meanwhile, an untreated case with the same high propensity score would receive a weight of $1/(1 - .9) = 10$. The propensity score of $p = .9$ implies that treated cases should outnumber control cases 9 to 1, because most cases with propensity scores that high should receive treatment. The nine treated cases would each receive weights of 1.111, becoming 10 'virtual' treatment cases. The one untreated case would receive a weight of 10, in effect also becoming 10 'virtual' cases. The number of treated and control cases with $p = .9$ in this virtual population would be balanced. And this will occur for every value of the propensity score. Applying the weights for ATE reweights the propensity score distributions of the treated and control cases, pushing both of them to the marginal distribution.

The weights for estimating the ATT can be obtained by multiplying Equation (5.1) by the propensity score, $p\,(D = 1|\,\mathbf{X})$, yielding

$$W_{\text{ATT}} = D + \frac{(1-D)\,p(D=1\,|\,\mathbf{X})}{1 - p(D=1\,|\,\mathbf{X})} \tag{5.3}$$

Like the ATT weights, one term of equation (5.3) drops out depending on D. When $D = 1$, W_{ATT} reduces to one; when $D = 0$, it becomes $p\,(D = 1|\,\mathbf{X})/[1-p\,(D = 1|\,\mathbf{X})]$. In other words, to estimate the ATT, all of the treated cases receive weights of one and all the control cases are weighted by the ratio of the probability of being treated to the probability of being untreated, conditional on the propensity score. This has the effect of creating a virtual population in which the propensity score distribution of control cases is pushed to the distribution of treated cases. This is precisely the situation illustrated in Figure 5.11.

The weighting scheme for estimating ATU follows similar logic. Equation (5.1) is multiplied by $1 - p(D = 1|\,\mathbf{X})$, yielding

$$W_{\text{ATU}} = \frac{D\,[1 - p(D=1\,|\,\mathbf{X})]}{p(D=1\,|\,\mathbf{X})} + (1-D) \tag{5.4}$$

Equation (5.4) switches between values depending on D, assigning a weight of one to untreated cases and a weight of $[1 - p(D = 1|\,\mathbf{X})] \,/\, p(D = 1|\,\mathbf{X})$ to the treated cases. Here, a virtual population is constructed in which the propensity score distribution of the untreated is left unchanged, and the treated cases are reweighted to push their distribution on the propensity scores to the distribution of untreated cases.

Table 5.4 summarises the IPTW for estimating the three treatment effects, with $p(D = 0|\,\mathbf{X})$ replacing $1 - p(D = 1|\,\mathbf{X})$ for clarity.

Table 5.4 IPTW equations for estimating ATE, ATT and ATU

Treatment effect	$D = 0$	$D = 1$
ATE	$1/p(D = 1|\mathbf{X})$	$1/p(D = 0|\mathbf{X})$
ATT	$p(D = 1|\mathbf{X})/p(D = 0|\mathbf{X})$	1
ATU	1	$p(D = 0|\mathbf{X})/p(D = 1|\mathbf{X})$

Note. IPTW = inverse probability of treatment weights; ATE = average treatment effect; ATT = average treatment effect for the treated; ATU = average treatment effect for the untreated.

Figure 5.13 illustrates the results of applying these three sets of weights to the data on the propensity score distributions. The target distribution in each panel is outlined.

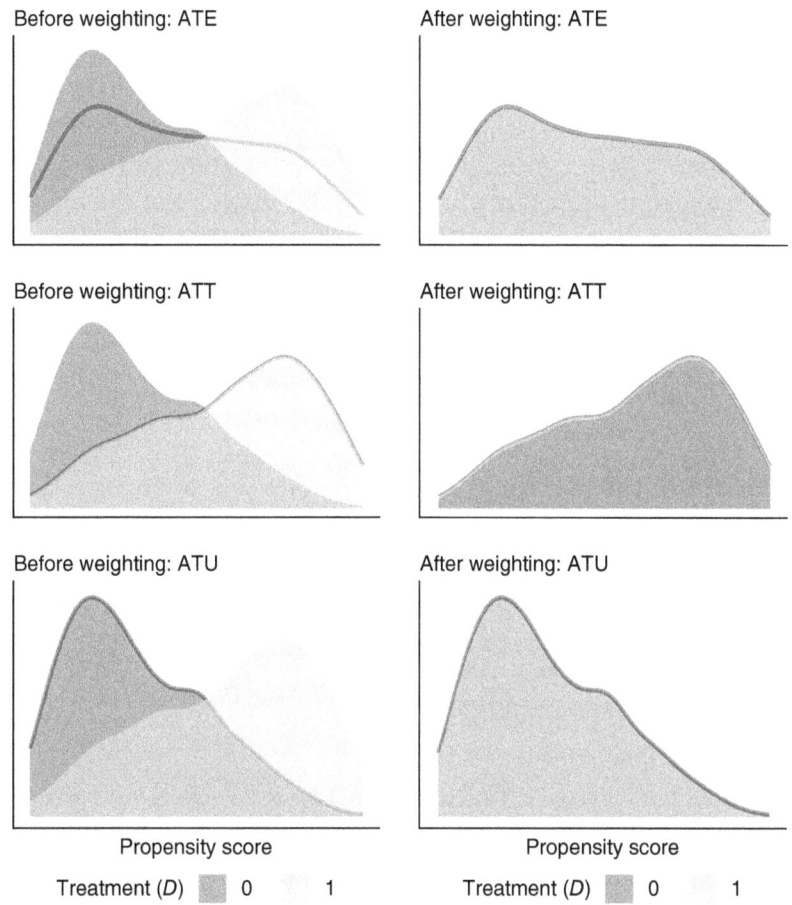

Figure 5.13 Using inverse probability of treatment weights to estimate causal effects. Left column: Before applying weights. Right column: After applying weights. First row: ATE. Second row: ATT. Third row: ATU. The target density is outlined. The target density for estimating ATE is the marginal density of propensity scores

Note. ATE = average treatment effect; ATT = average treatment effect for the treated; ATU = average treatment effect for the untreated.

Stabilised weights and truncated weights

Much of this book has been focused on reducing the bias of estimated causal effects, but researchers should also be mindful of their variance (represented by the standard error). In some cases, a slightly biased but low-variance estimator will tend to be closer to the true value than an unbiased but high-variance estimator. This is why statisticians often evaluate performance using the mean squared error (MSE), which contains contributions from both sources of error (James et al., 2013).

The variance of weighted estimators is affected by the magnitude of the weights. Large weights leads to a high-variance estimate. The more dissimilar the target distributions are from one another, the more extreme the weighting that is required to equalise them. Extreme weights mean that some observations are effectively 'counted' in the analysis many times, but although they may have the influence of multiple observations, they only contain the information of a single case. This mismatch between the influence a case has in the analysis and the information it contains reduces the precision of the estimated causal effects (Austin & Stuart, 2015).

The following output displays some selected percentiles of the ATE weights from the sample data.

```
##      50%       75%       90%       95%       99%      100%
##  1.455046  1.994656  3.096779  4.235763  7.582661  13.281683
```

It is clear that this is a substantially skewed distribution with a long right tail. In other words, there are a small number of relatively large weights. This small number of extreme weights harms the precision of estimation of the causal effect.

As a thought experiment, imagine a case in which the distribution of propensity scores for treated and untreated cases was identical before weighting. In this case, no action would be needed to move either distribution to the other (or to the marginal). All the weights would therefore be one. The efficiency of the resulting estimates would be maximal.

When the propensity score distributions are unequal across groups, the procedure requires larger and more variable weights to equalise the distributions. This reduces the precision of estimation. And the more dissimilar the propensity score distributions are, the more variable the weights must become in order to equate them. There is no free lunch. The effort required to equalise the distributions comes at a cost.

There are two approaches for limiting the variability of the weights. The first is to calculate *stabilised weights*, which are also sometimes called *corrected weights* (Guo & Fraser, 2015). Table 5.5 presents the equations for stabilised weights (Austin & Stuart, 2015). The equations for calculating stabilised weights are similar to the ones listed in Table 5.5, but the ones are replaced with the marginal probability of treatment or control, depending on which treatment effect is being estimated. The use of stabilised weights is recommended, as they yield more precise estimates without a corresponding increase in bias.

$$W_{\text{ATE, stabilised}} = \frac{(D)\,p(D=1)}{p(D=1|\mathbf{X})} + \frac{(1-D)\,p(D=0)}{1-p(D=1|\mathbf{X})} \tag{5.5}$$

Table 5.5 Stabilised IPTW equations for estimating ATE, ATT and ATU

Treatment effect	$D = 0$	$D = 1$
ATE	$p(D = 1)/p(D = 1|\mathbf{X})$	$p(D = 0)/p(D = 0|\mathbf{X})$
ATT	$p(D = 1|\mathbf{X})/p(D = 0|\mathbf{X})$	$p(D = 1)$
ATU	$p(D = 0)$	$p(D = 0|\mathbf{X})/p(D = 1|\mathbf{X})$

Note. IPTW = inverse probability of treatment weights; ATE = average treatment effect; ATT = average treatment effect for the treated; ATU = average treatment effect for the untreated.

Truncation (also called *trimming*) is another method for reducing the variance of the weights. It involves an explicit bias–variance trade-off that is often favourable in practice. The idea is to trim the long tail off the distribution of weights by setting all the weights above some high percentile to that value (Austin & Stuart, 2015). For example, if a weight of 5.1 is the 97th percentile, all weights above that number could be set to 5.1. Figure 5.14 shows how the application of truncated weights does not completely equalise the propensity score distributions of treated and control cases. The dissimilarity increases the potential bias in the treatment effect estimates but results in smaller standard errors and tighter confidence limits. Adaptive truncation of weights has been shown to be beneficial in some simulation work (Bembom & van der Laan, 2008).

Extreme weights are often indicative of poor common support. Rather than truncating the weights, it may be better to closely examine the distribution of propensity scores by group and impose a region of common support by discarding cases from the low-overlap regions. In this condition, extreme weights are not required to balance the groups. Discarding these non-overlapping cases reduces the external validity of the study but may increase the internal validity (Tipton, 2013).

Example of an analysis using propensity score weights

Performing an analysis with propensity score weights is quite simple in practice. The researcher needs only some method for estimating the propensity scores and for fitting weighted regression models. The weights themselves can readily be calculated with basic programming using the equations in Table 5.4 (unstabilised weights) or Table 5.5 (stabilised weights). If desired, the weights can be truncated. In R, the survey package provides the necessary tools for fitting weighted regression models via the svydesign() and svyglm() functions.

The R package twang ('*toolkit for weighting and analysis of non-equivalent groups*') is a user-friendly and powerful tool for estimating propensity scores, diagnosing residual imbalance using graphical and numeric presentations and automatically calculating ATE or ATT weights. (The ATU weights can be obtained by switching the numeric flags for treatment and control and requesting ATT weights.) The package contains its own customised implementation of the gbm package that was introduced earlier in this chapter for fitting boosted models: the optimum number of classifiers in the ensemble is determined by a covariate balance metric rather than minimising classification errors (e.g. Bernoulli deviance). Since the entire purpose of fitting the propensity score model is to generate covariate balance, this is a noteworthy feature.

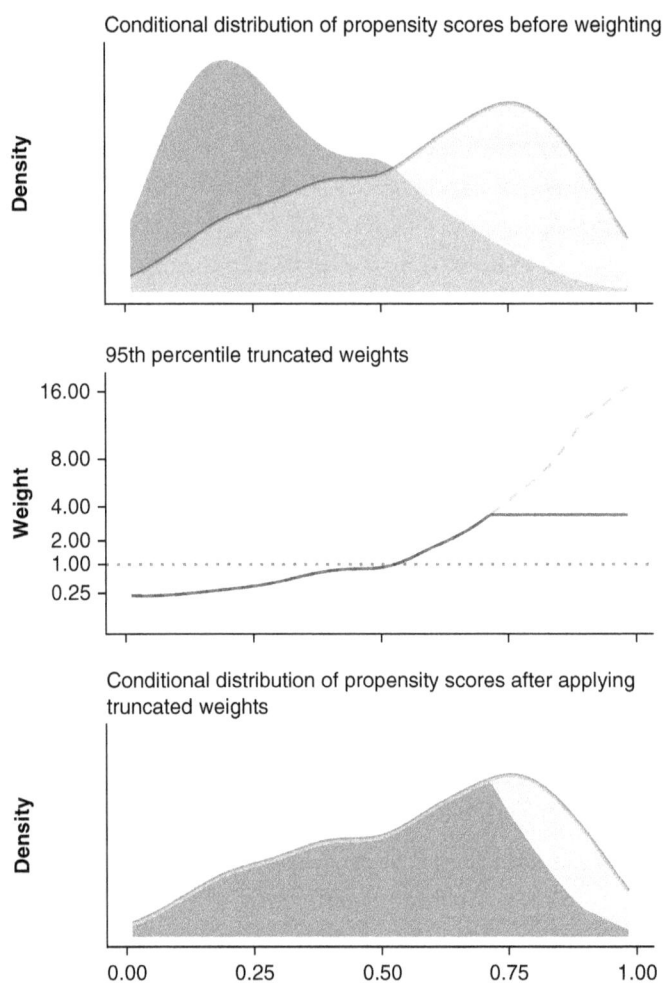

Figure 5.14 Illustration of truncated ATT weights. The weights have been truncated at the 95th percentile. Top panel: The unweighted distribution of propensity scores for the untreated (left) and treated (right) target distribution. Middle panel: The weights applied to the untreated cases. Truncated weights are depicted by a solid line; original weights by a dashed line. Bottom panel: The weighted distribution of propensity scores. After weighting, the distribution of propensity scores for the untreated resembles the distribution for the treated below the truncation point but deviates above it

Note. ATT = average treatment effect for the treated.

The ps() function in twang fits the boosted model and calculates the weights using a single line of code. The propensity scores and unstabilised weights can be extracted from the object returned by ps(). The bal.table() function produces diagnostic information regarding covariate balancing before and after the weights are applied. Its output, when run on the example data set, is given below:

```
## $unw
##    tx.mn tx.sd   ct.mn ct.sd  std.eff.sz   stat p    ks ks.pval
## A 0.410 0.973  -0.364 0.917       0.761 10.744 0 0.372   0.000
## B 0.161 1.064  -0.134 1.004       0.284  3.747 0 0.144   0.001
##
## $es.mean.ATE
##    tx.mn tx.sd   ct.mn ct.sd  std.eff.sz  stat     p    ks ks.pval
## A  0.040 0.966  0.091 0.997       0.133 1.361 0.174 0.066   0.658
## B -0.025 1.079 -0.067 1.001       0.041 0.441 0.660 0.047   0.939
```

The values listed in the $unw are the unweighted values. The means and standard deviations of each covariate in the treatment and control groups are listed. Of particular interest are the standardised effect sizes, presented in the std.eff.sz column. The pre-weighted effect sizes were $d = 0.761$ for variable A and $d = 0.284$ for variable B. Hypothesis test results for each of these variables are computed using a bootstrapping approach and are displayed in the ks.pval column; both A and B yield small and statistically significant p-values.

The values listed in $es.mean.ATE are the same balance statistics for the virtual population created by applying the weights. (The ATE in this object's name results from my having requested the ATE in the estimand= argument to ps(), while es.mean reflects my choice of the balance criteria used to select the optimal ensemble size.) After weighing, the effect sizes were $d = 0.133$ for A and $d = 0.041$ for B, and both of the p-values were non-significant. As an admittedly arbitrary heuristic, I like to see $d < 0.2$ after weighting, which this result obviously satisfies.

Plotting the object returned by the ps() function produces diagnostic plots. Though several are available, the balance plot shown in Figure 5.15 is among the most useful. This plot depicts the pre- and post-weighting effect sizes for the treatment versus control mean difference of all the covariates.

Running the summary() function on the object returned by ps() provides information for assessing the efficiency of the weighting process. The most important information in this output is the effective sample sizes for each group (ess.treat and ess.control) in the weighted condition.

```
##             n.treat n.ctrl ess.treat ess.ctrl    max.es     mean.es    max.ks
## unw            312    388  312.0000  388.000 0.7610431 0.52245570 0.37159662
## es.mean.ATE    312    388  205.4991  279.604 0.1332877 0.08706379 0.06553321
```

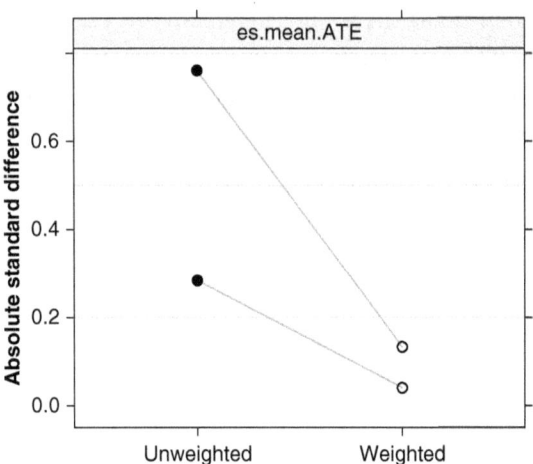

Figure 5.15 Balance plot for covariates after applying IPTW weights (ATE).

Note. IPTW = inverse probability of treatment; ATE = average treatment effect.

Though there are 312 treatment cases in the sample, there are effectively only 205.5 in the weighted sample. Similarly, there are 388 control cases, but effectively only 279.6 after weighting. This loss is a result of variance in the weights and results in reduced estimation efficiency in accordance with the diminished sample size.

Results using unstabilised weights

A weighted regression model is used to estimate the treatment effect using functions in R's survey package. The results are as follows:

```
##
## Call:
## svyglm(formula = Y ~ D, design = design.ps.ATE)
##
## Survey design:
## svydesign(ids = ~1, weights = ~weights.ATE, data = data)
##
## Coefficients:
##               Estimate Std. Error t value Pr(>|t|)
## (Intercept) -0.04778    0.07446  -0.642  0.52125
## D            0.45636    0.12164   3.752  0.00019 ***
## ---
## Signif. codes: 0 '***' 0.001 '**' 0.01 '*' 0.05 '.' 0.1 ' ' 1
##
## (Dispersion parameter for gaussian family taken to be 1.860452)
##
## Number of Fisher Scoring iterations: 2
```

The ATE estimated on the basis of the unstabilised weights is 0.456 with a 95% CI [0.218,0.695]. The true value of 0.3 is contained in the interval.

Results using truncated weights

In order to improve the efficiency of the resulting estimate, I truncated the ATE weights at their 95th percentile. This involved capping the weights at 4.236, which affected 35 of the cases, nearly all of which belonged to the treatment group. The truncation reduced the variance of the weights by 58%. The results of this analysis are shown below:

```
##
## Call:
## svyglm(formula = Y ~ D, design = design.ps.ATE.trunc)
##
## Survey design:
## svydesign(ids = ~1, weights = ~weights.ATE.trunc, data = data)
##
## Coefficients:
##               Estimate Std. Error t value Pr(>|t|)
## (Intercept) -0.07246    0.07273   -0.996   0.319
## D            0.45387    0.11390    3.985 7.46e-05 ***
## ---
## Signif. codes:  0 '***' 0.001 '**' 0.01 '*' 0.05 '.' 0.1 ' ' 1
##
## (Dispersion parameter for gaussian family taken to be 1.876409)
##
## Number of Fisher Scoring iterations: 2
```

The confidence interval narrowed slightly when the truncated weights were used. The estimated ATE with these weights was 0.454 with a 95% CI [0.231,0.677]. Like the previous result, the true value of 0.3 is contained in the interval.

Doubly robust estimation

The fundamental advantage of the methods of PSA presented in this chapter – subclassification, weighting and matching – is reduced model dependence compared to model-based statistical control. These procedures are dependent on fewer assumptions than linear statistical models and therefore work properly under a larger set of circumstances.

Regardless of which propensity score method is used, in reality there will almost always be significant residual imbalance or non-equivalence on the X variables even after conditioning on the propensity score. Doubly robust estimation applies statistical control to clean up this remaining imbalance (Funk et al., 2011) after conditioning on the propensity scores. The consequences of incorrect model specification depend on the degree of adjustment required. When the treatment and control groups are only slightly different with respect to some covariate, adjusting away the remainder of the imbalance can be successfully accomplished with a linear model even if the true functional form is non-linear. After all, a curve can be well-approximated by a line over a short distance. On small scales, the distinction between linear and non-linear functional form becomes negligible.

As an example of doubly robust estimation, the output below applies it to the propensity score weighting example from earlier in this chapter. The output shows that the weighted regression model now includes coefficients for A and B, reflecting their inclusion in the model.

```
## 
## Call:
## svyglm(formula = Y ~ D + A + B, design = design.ps.ATE)
## 
## Survey design:
## svydesign(ids = ~1, weights = ~weights.ATE, data = data)
## 
## Coefficients:
##               Estimate Std. Error t value Pr(>|t|)
## (Intercept) -0.05194    0.05307  -0.979    0.328
## D            0.42453    0.08373   5.070 5.09e-07 ***
## A            0.46755    0.05165   9.051  < 2e-16 ***
## B           -0.69234    0.04916 -14.082  < 2e-16 ***
## ---
## Signif. codes: 0 '***' 0.001 '**' 0.01 '*' 0.05 '.' 0.1 ' ' 1
## 
## (Dispersion parameter for gaussian family taken to be 0.8393258)
## 
## Number of Fisher Scoring iterations: 2
```

The ATE estimated with the doubly robust weighted analysis is 0.425 with a 95% CI [0.26,0.589]. For context, compare this result to the weighted result that did not incorporate the additional regression adjustment: ATE = 0.456, 95% CI [0.218, 0.695].

The true value of 0.3 is contained in both the intervals, but the doubly robust model's point estimate is closer to the population value of the treatment effect, and the estimate is substantially more precise as indicated by the narrower confidence interval.

Further Reading

Adelson, J. L. (2013). Educational research with real-world data: Reducing selection bias with propensity scores. *Practical Assessment, Research, and Evaluation, 18*(15), 2–11.
This article introduces propensity score stratification to education researchers and uses it to analyse data from the Early Childhood Longitudinal Study.

Austin, P. C. (2011). An introduction to propensity score methods for reducing the effects of confounding in observational studies. *Multivariate Behavioral Research, 46*(3), 399–424. https://doi.org/10.1080/00273171.2011.568786
This article describes propensity score techniques as an alternative to randomised experiments and introduces three of the PSA techniques considered in this chapter (matching, stratification, and weighting) as well as simply using the propensity scores as a covariate in a linear model. The articles closes by comparing propensity score methods with standard regression adjustment.

Austin, P. C., & Stuart, E. A. (2015). Moving towards best practice when using inverse probability of treatment weighting (IPTW) using the propensity score to estimate causal treatment effects in observational studies. *Statistics in Medicine, 34*(28), 3661–3679. https://doi.org/10.1002/sim.6607
This article focuses on the use of propensity scores to calculate inverse probability of treatment weights for balancing the sample on confounders. The article describes a number of statistical and graphical techniques for assessing the quality of balance after weighting.

Funk, M. J., Westreich, W., Weisen C., Stürmer, T., Brookhart, M. A., & Davidian, M. (2011). Doubly robust estimation of causal effects. *American Journal of Epidemiology, 173*(7), 761–767. https://doi.org/10.1093/aje/kwq439
This article introduces the concept of doubly-robust estimation, showing that unbiased estimation of causal effects occurs when either the propensity score or the regression adjustment models are correctly specified. They assess the performance of doubly-robust estimation using a Monte Carlo simulation.

Guo, S., & Fraser, M. W. (2015). *Propensity score analysis: Statistical methods and applications* (2nd ed.). Sage.
Guo and Frasier's textbook is, at the time of this writing, the most complete and comprehensive text on propensity score techniques.

Stuart, E. A. (2010). Matching methods for causal inference: A review and a look forward. *Statistical Science, 25*(1), 1–21. https://doi.org/10.1214/09-STS313

This article provides a gentle introduction to Rubin's causal model and then discusses various matching algorithms in detail. It closes by discussing topics that are rarely considered in the introductory literature on propensity score matching, such as missing data, multiple-valued treatments, and how to choose between the various methods.

6

INSTRUMENTAL VARIABLE ANALYSIS

Chapter Overview

All of the methods of causal inference discussed thus far in this book have been conditioning methods. These techniques require the identification and precise measurement of the variables necessary to achieve ignorability or block all back-door paths from the exposure to the response variable.

Certainly DAGs help clarify one's reasoning about the web of causal relationships enveloping a research project's central variables, facilitate communication of such theories to research consumers and identify the set (or sets) of variables that must be conditioned on in order to estimate causal effects. But sometimes, analysis of a DAG may reveal that all sufficient adjustment sets contain unobservables – variables that were not or cannot be measured – and that no conditioning method is sufficient for estimating the causal effect. In other situations, the variables may have been observed, but they were measured poorly. And perhaps most commonly, researchers may simply harbour deep scepticism about the veracity of their proposed DAG, even when the graph passes tests of its implied independencies and conforms to accepted theory. Perhaps key variables were not included, interactions were omitted or the directions of some arrows were incorrect. Figure 6.1 displays two DAGs that differ in the direction of only one arrow. This single distinction has large implications for the implied sufficient adjustment sets. Variable Z *must* be conditioned on according to the left panel but must *not* be in the right panel. This example is rather trivial, but such errors are easy to make and difficult to detect in large, realistic DAGs containing dozens of variables.

Figure 6.1 Two examples of DAGs. Left panel: Z as a confounder. The analysis must condition on it. Right panel: Z as a mediator. The analysis must not condition on it. Reversing one arrow in the DAG dramatically alters the composition of the adjustment set

Note. DAG = directed acyclic graph.

Ignorability (or fulfilment of the back-door criterion) is a strong and fundamentally untestable assumption, and it pays to maintain a high degree of scepticism about whether it has actually been achieved. The careful construction and assessment of DAGs, along with the use of PSA, can clarify assumptions and increase the robustness of the analysis, but these techniques are no panacea. Sometimes the most productive course of action is to accept that one does not have the necessary information (or available measurements) to productively deploy a conditioning strategy. In that case, researchers have essentially three remaining options for estimating causal effects:

- If treatment assignment can be manipulated practically, ethically, at a reasonable cost and within an acceptable time frame, the researcher can run a randomised experiment.
- If the researcher cannot randomise cases to treatment conditions, but is aware of a process that assigns subjects to treatments based on their level of some continuous variable, the researcher may be able to employ a regression discontinuity design (RDD), which is the subject of Chapter 7.
- If an instrumental variable (IV) can be identified, researchers can use instrumental variables analysis (IVA) to estimate the causal effect.

A shortcoming of randomisation and the RDD is that the exposure variable (e.g. the 'treatment') must be categorical in nature. This limitation does not apply to IVA; it can be used regardless of the nature of the treatment variable. This chapter introduces IVA and examines it from a DAG perspective, which clarifies its motivation, rationale and vulnerabilities.

Endogeneity and bias

The central challenge to the estimation of causal effects is confounding bias. IV methods were developed in economics and are fundamental to research in that field, which has developed a somewhat different vocabulary and framework for discussing these issues that one finds in other branches of the social sciences (Greenland, 2000). In some cases, econometricians use slightly different definitions for terms that are also central to the DAG and RCM literatures.

In the IVA world view, the problem of confounding is understood in terms of statistical models and their assumptions – specifically, how confounding violates those assumptions (Maddala & Lahiri, 1992). This stands in opposition to the DAG and RCM perspectives, which situate the problem of causal inference at a higher level of abstraction. For example, the issue of confounding in the DAG framework is agnostic to the identity or form of any candidate statistical model that might be used to analyse the data. The same is true in Rubin's framework, in which confounding creates a difference between observable values and their counterfactual counterparts (Angrist et al., 1996). Under both perspectives, confounding bias exists regardless of whether researchers use a parametric statistical model or a completely non-parametric method to analyse the data. The IVA framework situates the confounding problem at a more pragmatic level – the level of the statistical models that will be employed to estimate the causal effect of interest. In this framework, the problem with confounding is that it violates model assumptions and therefore results in biased and inconsistent estimates of causal effects.

The term *endogenous* was introduced in Chapter 3, where it was defined as a variable in a DAG that is caused by another variable in the graph. An endogenous variable has an arrow directed at it. The term has a slightly different definition in the IV framework: a predictor variable that is correlated with a statistical model's prediction errors is endogenous (Greenland, 2000). A core assumption of linear regression and related techniques is that the prediction errors are uncorrelated with the predictor variables. **Endogeneity** is a violation of this assumption resulting from confounding, non-random measurement error, or even simultaneity – in which the current value of X is caused by a previous value of Y. This forbidden correlation is the root cause of the estimation bias that prevents the model's results from capturing the magnitude of the causal relationship between X and Y. This bias is referred to as *omitted variable bias* in the statistics and econometrics literature (Berk, 2004).

From a DAG point of view, this problematic correlation between the exposure variable and the model's errors exists precisely because of endogeneity in the actual causal system as represented in the graph – that the exposure variable shares a common cause with the outcome (Pearl, 2009). If the exposure variable was exogenous in this larger sense, it would have no in-graph causes by definition, and would therefore be unconfounded, and statistical models of the $X \rightarrow Y$ relationship would not be endogenous in the IV sense of the term.

It is very important to distinguish between *errors* and *residuals*. Though these terms are often used interchangeably, they are different concepts. Errors are a property of the population and are the differences between the observed values of Y and their predicted values according to the population regression model. Equation (6.1) defines errors in the simple regression context.

$$\varepsilon_i = Y_i - [\beta_0 + \beta_1(X_i)] \qquad \text{(errors)} \qquad (6.1)$$

By contrast, residuals are a property of the sample. They are the deviations between observed Y and predicted values from an estimated regression model. Equation (6.2) defines residuals, again in the context of simple regression.

$$e_i = Y_i - [b_0 + b_1(X_i)] \qquad \text{(residuals)} \qquad (6.2)$$

The residuals are defined in terms of the regression parameter estimates, b, while the errors are defined with respect to the true regression parameters, β. The residuals (e) are statistical estimates of the errors (ε). When endogeneity exists, the estimated regression coefficients are biased, so they systematically deviate from their population counterparts. Since the residuals are a function of the regression coefficients, endogeneity causes them to systematically deviate from the errors. Both the coefficients and the residuals are disturbed by endogeneity.

The residuals are constructed such that they are always orthogonal to, and therefore uncorrelated with, the predictor variables (Fox, 1997). This implies that one cannot test for endogeneity by examining correlations of the residuals against the **X** variables. Even when endogeneity exists, the model-estimated residuals will be uncorrelated with **X**. The actual errors will be correlated with **X**. In fact, it is the biased estimation of the regression coefficients that allows the residuals to be uncorrelated with the predictor variables.

Figure 6.2 displays the difference between residuals and errors. These scatter plots were generated from simulated data in which the $X \rightarrow Y$ relationship is disturbed by a confounder. The *residuals* displayed on the y-axis of the left panel were calculated by fitting a simple linear regression model of the form $Y_i = b_0 + b_1(X_i) + e_i$ to the data. These residuals are plotted against the X variable. This model is misspecified because it does not include the confounder, so its estimate of the causal effect of X on Y is biased and inconsistent. But, as shown in the scatter plot, there is no correlation between X and the model residuals.

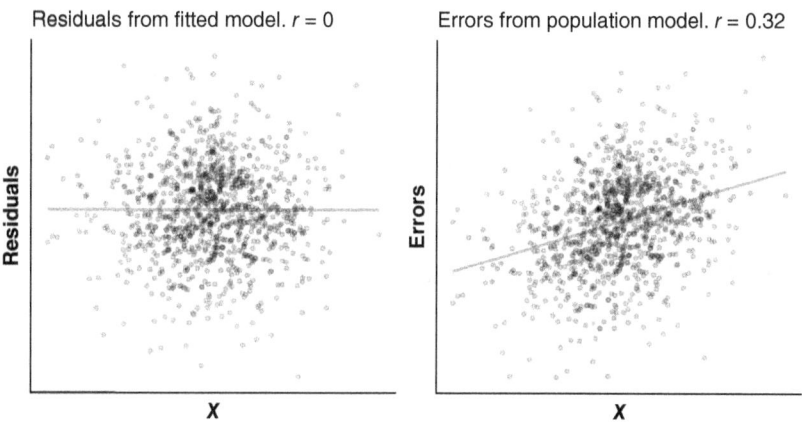

Figure 6.2 Residuals versus errors under endogeneity. Left panel: Scatter plot of X versus the residuals from the fitted model shows no correlation between them. Right panel: Scatter plot of X versus the errors derived from the true population model. This panel reveals that X is correlated with the errors

The right panel of Figure 6.2 is a scatter plot of the model *errors* versus X. This is a graphical depiction of endogeneity. This endogeneity manifests as a badly biased estimate of the treatment effect. I could calculate the errors for this plot since I know the true population values of parameters β_0 and β_1, which I had to specify in order to simulate the data. The errors are correlated with the X variable, indicating that the X variable is endogenous. Unfortunately, because the errors are unobservable in real-life contexts in which the true regression coefficients are unknown, this method has no diagnostic utility. A model's residuals will always be uncorrelated with the predictor variables even if its errors are not (Fox, 1997).

Defining instrumental variables

An **instrument** (or **instrumental variable**) was defined in Chapter 3 as a variable that has a causal effect on the exposure and whose effect on the response is completely mediated by the exposure, such that the true DAG has the form $I \rightarrow X \rightarrow Y$, where I is the instrument. This means that the instrument cannot have any direct effect on the response, nor can it be a confounder or a proxy confounder of any path connecting the exposure and the response. Its entire influence on Y must pass through X.

IVs are sometimes defined in statistical terms rather than in terms of their causal relations with the exposure and outcome variables. Bollen (2012) defined IVs as having two properties. First, they are correlated with the exposure variable X. Second, they must not be correlated with the errors of a model predicting Y from X. This second condition implies that the instrument I and the outcome Y must be independent after conditioning on X. These two conditions are satisfied by any variable that qualifies as an instrument in the DAG framework. If I causes X, they will be correlated, and if X completely mediates the $I \rightarrow Y$ relationship, these variables are independent conditional on X. However, this statistical criterion is slightly weaker because it only requires that the instrument be correlated with X, not necessarily a cause of X.

This fundamental assumption of IVA – that the instrument is only related to **Y** through **X** – is called the exclusion restriction (Angrist et al., 1996). In many research contexts, the exclusion restriction can be more plausible than conditional ignorability.

The exclusion restriction is only partially testable (Kitagawa, 2015). Whether it is met in practice is always an assumption. Obviously, one has reason to reject it if a partial correlation or linear regression reveals a non-zero statistical relationship between the instrument and the response after conditioning on the exposure or if other exogeneity diagnostics reveal evidence of a problem (Bascle, 2008). But passing diagnostic tests does not prove that the putative instrument actually is one. Any method for testing for independence is conditioned on its own assumptions, which may not be met. For example, uncorrelated does not imply independence, only *linear* independence. A variable might pass this diagnostic test without actually being an instrument if it is connected to Y through multiple mediating pathways of opposite sign whose net effect cancels out. Or, most prosaically, the diagnostic test might generate a false-negative (e.g. Type II) error due to low statistical power or bad luck.

Figure 6.3 displays a DAG with an instrument (I) and an unobservable confounder (U). The causal effect of X on Y cannot be identified via conditioning because of the unobservable confounder U. But the instrument can be exploited to make it identifiable.

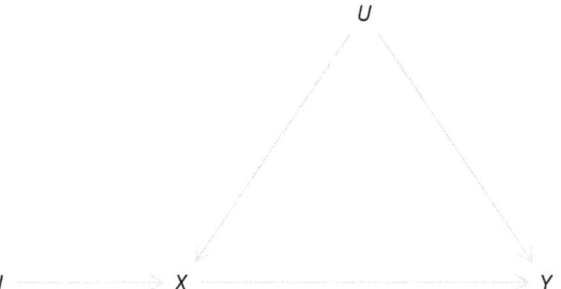

Figure 6.3 Example of a DAG with an instrument. X is the exposure, Y is the response, U is an unobservable confounder and I is an instrument

Note. DAG = directed acyclic graph.

There are also conditional instruments, as shown in Figure 6.4 (Baiocchi et al., 2014). Here, variable C is a confounder of the relationship between I and Y. Therefore, I is not an instrument, but it becomes one conditional on C. One should not generally use a conditional instrument to do IVA; if any confounder exists, it is likely that there are others. Also, it is unlikely that the confounder of the $I \rightarrow Y$ path is perfectly measured, so there will be some residual confounding even after conditioning on C. The IV estimator is strongly biased if there is even a small violation of the exclusion restriction (Pearl, 1993). There is one notable exception to this advice: the fuzzy RDD, which will be discussed in Chapter 7. In that case, the fundamental design guarantees the existence of a conditional instrument.

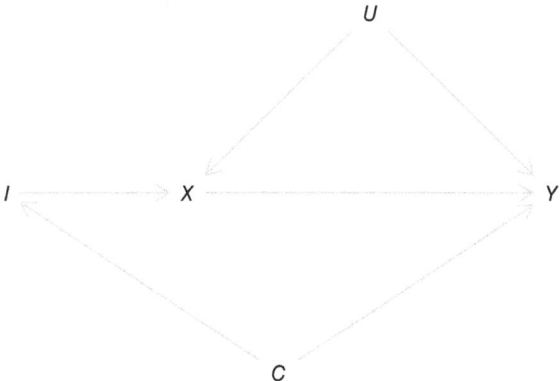

Figure 6.4 Variable I is not an instrument, but it becomes one conditional on variable C

A direct causal path from I to X is sufficient but not necessary for establishing a variable as an instrument. The requirement is that the instrument is merely *correlated* with X while obeying the exclusion restriction. This correlation could arise as a result of confounding of the I to X path or unobserved mediators

between *I* and *X*. Figure 6.5 shows such a scenario. In the left panel of the figure, the instrument *I* has no causal effect on *X*. It is correlated with *X* due to the
influence of unobservable confounder *U2*. In the right panel, there is an unobserved mediator *M* between *I* and *X*. In either of these conditions, variable *I* still
fulfils the requirements of being an instrument and can be used to identify the
X → *Y* causal effect.

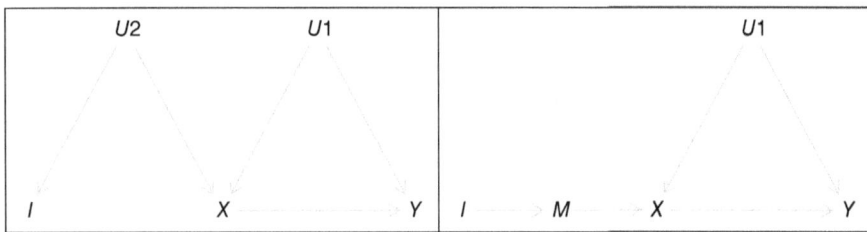

Figure 6.5   Example of DAGs in which instruments do not have a direct, contiguous causal
connection to the exposure. Left panel: Unobservable confounder *U2* creates a correlation
between *I* and *X*. Right panel: Mediator *M* transmits the causal effect from *I* to *X*

Note. DAG = directed acyclic graph.

I simulated a data set from this DAG with $n = 1000$ to use in the examples in this
chapter. The true treatment effect of $X \rightarrow Y$ is 1.0. This data set is available for download from this book's companion website. Even though *U* is supposed to represent
an unobservable confounder, I nonetheless included it in the simulated data set. This
enables the results obtained from IVs methods to be compared with the results from
conditioning methods. But as a means of motivating IVA, one should pretend that *U*
is unavailable.

Fitting a simple regression model $Y_i = b_0 + b_1(X_i) + e_i$ to the data produces the
following output.

```
##
## Call:
## lm(formula = Y ~ X, data = data)
##
## Residuals:
##       Min       1Q    Median       3Q      Max
## -25.0763   -4.4373   -0.0399   4.7108   23.0404
##
## Coefficients:
##                 Estimate Std. Error t value Pr(>|t|)
## (Intercept) -0.09150      0.22735  -0.402     0.687
## X            2.00931      0.09447  21.270    <2e-16 ***
## ---
## Signif. codes: 0 '***' 0.001 '**' 0.01 '*' 0.05 '.' 0.1 ' ' 1
```

```
##
## Residual standard error: 7.189 on 998 degrees of freedom
## Multiple R-squared: 0.3119,  Adjusted R-squared:  0.3112
## F-statistic: 452.4 on 1 and 998 DF,  p-value: < 2.2e-16
```

The estimated effect of X is 2.009, which is much too big compared to the true value of 1.0. Further, the true value was far from being contained in the confidence interval around the estimate, which is 2.009 (0.094), 95% CI [1.824, 2.194].

If U had been a measured variable, the analysis could have simply conditioned on it using any of the strategies outlined in Chapter 2. Below are the results from a regression analysis in which U has miraculously been included in the model.

```
##
## Call:
## lm(formula = Y ~ X + U, data = data)
##
## Residuals:
##      Min       1Q   Median       3Q      Max
## -19.7026  -3.7149   0.1242   3.6740  17.1255
##
## Coefficients:
##               Estimate Std. Error t value Pr(>|t|)
## (Intercept)  -0.01976    0.17404  -0.114     0.91
## X             1.03715    0.08103  12.800   <2e-16 ***
## U             5.01160    0.18855  26.580   <2e-16 ***
## ---
## Signif. codes: 0 '***' 0.001 '**' 0.01 '*' 0.05 '.' 0.1 ' ' 1
##
## Residual standard error: 5.503 on 997 degrees of freedom
## Multiple R-squared: 0.5973,  Adjusted R-squared:  0.5965
## F-statistic: 739.4 on 2 and 997 DF,  p-value: < 2.2e-16
```

Incorporating U into the model does result in an unbiased estimate of the causal effect of X. The estimated effect of X is 1.037, which is very close to the true value of 1.0. But this approach is obviously impossible when confounders are *unobservable*. The standard error of the effect of X (0.081) from this model is of particular interest. It will serve as the benchmark for comparison of the precision of results estimated using IVA.

Finding an instrument

Identifying a valid instrument can be quite challenging. For example, imagine that one wants to estimate the causal effect of cigarette smoking on respiratory function.

It is easy to think of many factors that are related to the probability of smoking: socio-economic status, education level, health consciousness, living with a smoker or having a work environment conducive to smoking, among many others. But for these to be instruments, one must be able to rule out, with a high degree of confidence, the possibility that these factors also cause the outcome independent of smoking – even indirectly through mediators. The possibility that these prospective instruments are related to the response through confounders must also be eliminated with a high degree of certainty.

Policy variables or sudden historical events are often used as instruments. For example, cigarette taxes can fluctuate widely across cities, counties or states. Rates of cigarette smoking do respond to price – many more people will smoke at $2 per pack than they would at $8 per pack. And the exclusion restriction is at least plausible – it is reasonable to assume that cigarette taxes can have no direct causal effect on lung function that is not mediated by cigarette use. So the tax rate could conceivably be an instrument. However, if smokers choose to relocate to low-tax areas in order to enjoy a lower cost of cigarettes, the tax rate would not be an instrument. Perhaps more plausibly, if localities with a high density of smokers elect politicians who enact low cigarette taxes, the variable would not be an instrument. In fact, the price of cigarettes has been used as instrument in several studies, including Evans and Ringel (1999), Leigh and Schembri (2004) and Mullahy (1997).

Natural disasters often make excellent instruments. For example, imagine that a tobacco blight or a widespread drought dramatically reduces the tobacco yield one year, resulting in a rapid price increase due to tight supply, or perhaps even supply shortages. This type of event imposes a nearly discontinuous change in the exposure variable. At the same time, it is quite difficult to imagine how a tobacco blight or a drought could affect respiratory health except by changing smoking behaviour. Miguel et al. (2004) used variation in rainfall as an instrument to study the relationship between economic growth and civil conflict in African nations. Another potential instrument could be geographic distance from a site. For example, the relative geographic distance from home to the zoned public school versus the nearest private school could be an instrument for private school attendance. Distance from the nearest urgent care clinic could be an instrument for emergency room use. Lu (1999) used distance from a mental health provider as an instrument to study the effectiveness of mental health care on productivity.

The best instrument is experimental randomisation (Heckman, 1995). In a successful experiment, the randomisation process completely controls treatment assignment without exception. In this case, the instrument is perfect, and specialised IVA is unnecessary. Figure 6.6 shows how randomised allocation to the exposure variable (X) functions as an instrument.

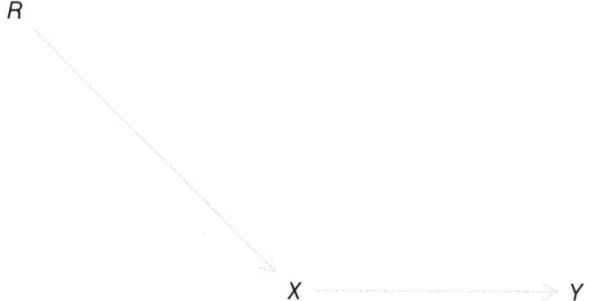

Figure 6.6 Example of a DAG where the exposure variable is randomly assigned. *R* is the randomisation process, *X* is the exposure and *Y* is the response. The randomisation process is an instrument

Note. DAG = directed acyclic graph.

However, there are many examples of 'broken' randomised studies in which compliance with the treatment assignment is imperfect (Barnard et al., 2003). Imagine an exercise science study in which the treatment group is asked to perform intense exercise sessions three times per week. Some of the treatment subjects will do this inconsistently, others may not do it at all. And some people in the control group are probably already participating in similar exercise programmes. In this case, the randomisation results in 'treatment encouragement' rather than treatment assignment since the research team cannot force compliance (West et al., 2008). In this case, the randomised encouragement can be viewed as an instrument. In this case, any placebo effect from simply being randomised to the treatment condition (but not actually receiving or participating in the treatment) would violate the exclusion restriction of IVA.

The literature is rife with examples of clever instruments. Angrist and Krueger (2001) provide a table (p. 82) with examples of instruments and the causal effects that they are intended to identify. Rassen et al. (2009) provided examples of instruments used in clinical epidemiology. Stukel et al. (2007) employed the regional cardiac cauterisation rate as an instrument to study the effect of invasive cardiac interventions on heart attack survival. Kim et al. (2011) studied the effect of contextual social capital on health in 64 countries using population density and measures of corruption as instruments. More generally, Mendelian randomisation is an approach to studying the causes of disease that uses genotypes as instruments (Didelez & Sheehan, 2007). There have been some excellent discussions of instrument identification online on the *Quora* (www.quora.com) and *CrossValidated* websites (www.stats.stackexchange.com), which can be found via a web search.

The two-stage least squares estimator

IVA is a technique for estimating causal effects from observational designs using instruments. The instrument provides another means of approximating an experiment from an observational study. The crux of the technique is to separate the variance of the exposure variable into two parts (Greenland, 2000). One component is solely a function of the instrument. This part is unconfounded because the instrument itself is unconfounded. In most cases, this will only comprise a small amount of the total variability of the exposure, and it is guaranteed by the exclusion restriction to be free of any variance shared with confounders. The other component of the exposure, which is discarded, contains 'everything else', including the confounding. In IVA, the full exposure variable is replaced by its unconfounded portion. Measuring the association between this subcomponent of the exposure and the response variable produces an asymptotically unbiased and consistent estimate of the treatment effect.

IVA is not guaranteed to produce unbiased estimates in small samples; the estimator is *consistent* but not unbiased (Martens et al., 2006). In practice, people do not seem to worry about this property very much, but it does provide a rationale for limiting the use of IVA to large samples. The arguably more important reason for large samples is that the estimates produced using IVA tend to have poor precision, as will be discussed shortly.

A dominant method for implementing IVA is known as *two-stage least squares* (2SLS) estimation (Angrist & Krueger, 2001). Alternative IV estimators are available and were described and reviewed by Burgess et al. (2017), but this chapter focuses on 2SLS because it is commonly implemented in software, has good performance and has a clear intuitive interpretation. Let the exposure variable be denoted X, the outcome Y and the instrument I. The stages of 2SLS estimation are as follows:

1 Regress the exposure variable on the instrument(s) with a model of the form $X_i = a_0 + a_1(I_i) + e_i$. Save the predicted (or fitted) values of $X(\hat{X})$ from this model.
2 Regress the outcome on the predicted X values with a model of the form $Y_i = b_0 + b_1(\hat{X}) + \xi_i$.

Parameter b_1 from the second model provides an asymptotically unbiased and consistent estimate of the causal effect if the assumptions are satisfied, the most important of which is the exclusion restriction. Modern software implements the two stages simultaneously in order to correctly estimate the standard errors and R^2 statistic from the second model. Implementing these steps as separate regression models means that the predicted values of X generated from the stage 1 model are presented to the stage 2 model as though they were known rather than estimated with their own uncertainty (Wooldridge, 2016). However, the explicit two-step approach has pedagogical value by revealing intermediate results that are hidden by a single-stage implementation of 2SLS. The point estimate for the $X \to Y$ causal effect is identical under either approach.

Step 1: Obtain the predicted values of the exposure variable

The purged component of the exposure (X) can be estimated by saving the predicted values, $\hat{X} = E(X \,|\, I)$, from the following regression model. If more than one instrument is available, they should all be included. (In practice, it is difficult enough to find a single instrument.)

$$X_i = b_0 + b_1 (I_i) + e_i \qquad (6.3)$$

The lm() output from this model is given below.

```
##
## Call:
## lm(formula = X ~ I, data = data)
##
## Residuals:
##      Min      1Q  Median      3Q     Max
## -8.3830 -1.5102 -0.0115  1.5077  7.6037
##
## Coefficients:
##               Estimate Std. Error t value Pr(>|t|)
## (Intercept)    0.02266    0.07411   0.306     0.76
## I              0.53778    0.07128   7.544 1.02e-13 ***
## ---
## Signif. codes:  0 '***' 0.001 '**' 0.01 '*' 0.05 '.' 0.1 ' ' 1
##
## Residual standard error: 2.343 on 998 degrees of freedom
## Multiple R-squared: 0.05395, Adjusted R-squared:  0.05301
## F-statistic: 56.92 on 1 and 998 DF, p-value: 1.022e-13
```

Note that the instrument has a positive and statistically significant relationship with the exposure variable, as well as the model's R^2 value of 0.054. This R^2 value measures the 'strength' of the instrument. Stronger instruments are desirable because they increase the precision and statistical power of the results from the stage 2 model. That is why multiple instruments should be included in this model if more than one has been identified. Moreover, the use of a weak instrument cannot remove all of the confounding bias and also results in a highly impre-cise (high variance) estimate of the causal effect (Angrist & Krueger, 2001). In the limiting case, the relationship between I and X is zero, and all the predicted values from the model are the mean of X. Staiger and Stock (1997) suggested that the first-stage model should have an F statistic of at least 10.0. This model has F (1,998) = 56.92 and is well above this criterion.

Figure 6.7 displays a scatter plot of the instrument (I) versus the actual values of the exposure (X) from the DAG in Figure 6.3 as well as the regression solution from Equation (6.3).

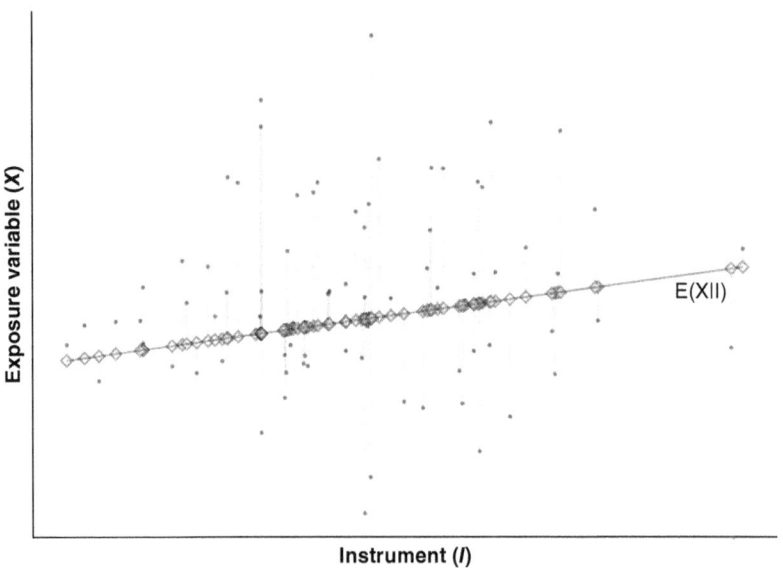

Figure 6.7 Predicted versus actual values of the exposure variable

It is clear from the figure that the variance of the predicted exposure variable is much smaller, only 5.4% of the original amount. Note that this is the model R^2 value from the regression output above. Purging the confounding from X has come at a high price. Almost all of its variance has been discarded.

Step 2: Estimate the treatment effect

In the stage 2 model, the treatment effect is estimated by regressing the response vari-able Y on the predicted values of the exposure ($\hat{X}$) generated from the stage 1 model. The lm( ) output is given below.

```
##
## Call:
## lm(formula = Y ~ pred.X, data = data)
##
## Residuals:
##      Min      1Q  Median      3Q     Max
## -29.213  -5.631   0.468   5.920  31.340
##
## Coefficients:
##                Estimate Std. Error t value Pr(>|t|)
## (Intercept) -0.08159    0.27320   -0.299  0.76528
## pred.X       1.29333    0.48857    2.647  0.00825 **
## ---
## Signif. codes: 0 '***' 0.001 '**' 0.01 '*' 0.05 '.' 0.1 ' ' 1
##
## Residual standard error: 8.637 on 998 degrees of freedom
## Multiple R-squared: 0.006973, Adjusted R-squared:  0.005977
## F-statistic: 7.007 on 1 and 998 DF, p-value: 0.008245
```

Simultaneous two-stage least squares

In R, the ivreg() function is useful for performing IVA using 2SLS. This function is part of the AER package (short for *Applied Econometrics With R*). (*Note:* The tsls() function from the sem package is also available.) Setting the diagnostic argument to TRUE when running summary() on the ivreg() results generates additional diagnostic information, including the *F* statistic from the stage 1 model.

The motivation for using a simultaneous approach is to obtain correct standard errors and model R^2 = statistic. It is easier to implement than the two-step regression approach and requires only a single line of code to invoke. The ivreg() output is given below.

```
##
## Call:
## ivreg(formula = Y ~ X | I, data = data)
##
## Residuals:
##      Min      1Q  Median      3Q     Max
## -23.761  -5.011   0.228   4.880  23.572
##
## Coefficients:
##                Estimate Std. Error t value Pr(>|t|)
## (Intercept) -0.08159    0.23387   -0.349  0.72726
## X            1.29333    0.41824    3.092  0.00204 **
```

```
##
## Diagnostic tests:
##                  df1 df2 statistic  p-value
## Weak instruments   1 998    56.917 1.02e-13 ***
## Wu-Hausman         1 997     3.283   0.0703 .
## Sargan             0  NA        NA       NA
## ---
## Signif. codes:  0 '***' 0.001 '**' 0.01 '*' 0.05 '.' 0.1 ' ' 1
##
## Residual standard error: 7.393 on 998 degrees of freedom
## Multiple R-Squared: 0.2723, Adjusted R-squared:  0.2716
## Wald test: 9.563 on 1 and 998 DF, p-value: 0.002041
```

The simultaneous 2SLS estimate of the causal effect is 1.293, which is identical to the value obtained in step 2 of the previous approach and reasonably close to the true value of 1.0. Moreover, the true value is contained in the 95% CI [0.474, 2.113]. The F-statistic from the stage model is reported as the Weak instruments test; it should be greater than 10.0 as per Staiger and Stock (1997).

There are two major differences between the ivreg result and what was obtained from using the manual 2SLS process. First, the standard error of X is slightly different using ivreg(). This is a good thing, because the manual two-step 2SLS procedure does not estimate the correct standard error. The manual 2SLS process treated $\hat{X}$ as though it was a known value, when in fact it was estimated with uncertainty in the first ordinary least squares regression model. Using a function like ivreg() or tsls() corrects this problem. Second, the ivreg result corrects the model R^2 value quite dramatically.

Notice that the standard error for the effect of X in this model (0.418) is much larger than in the linear regression model that conditioned on U (0.081); it is 5.16× larger. This is a highly undesirable but unavoidable issue in IVA. The precision of estimation (and therefore, statistical power) in regression is a function of the variance of the predictor variable. The effect of high-variance predictors can be estimated with more precision because the model 'sees' a wider range of values for X and how Y changes in response.

In IVA, the variance of the exposure variable is purposefully being reduced in order to purge the confounding from it. Here is another example of the bias–variance trade-off that has been discussed throughout this book. IVA trades precision for bias reduction.

It is always a good idea to include competing exposure variables (see Chapter 3) in regression models. Although these variables are not required to achieve unbiased estimates, they do increase the precision and statistical power of the estimates by absorbing some of the residual variance of the outcome. This becomes even more important in IVA because their precision is typically quite poor, especially when weak instruments are used. Anything that can be done to address this situation is worthwhile. Very large samples can overwhelm the baseline inefficiency of IVA and estimate causal effects with a useful level of precision. But high variance may result in a typical result being as far away (or farther) from the true value as a high-precision but biased estimate from a naive statistical analysis.

Even though the instrument protects against all confounding, both observed and unobservable, it is still worthwhile to measure as many would-be confounders and competing exposure variables as possible to include in the stage 2 model. Including these increases the precision of the estimated causal effect of the exposure. Below is the ivreg() output when competing exposure variable A is included in the model.

```
##
## Call:
## ivreg(formula = Y ~ X + A | I + A, data = data)
##
## Residuals:
##       Min       1Q    Median       3Q      Max
## -18.30686  -4.06584   0.08623   4.04599  19.92331
##
## Coefficients:
##               Estimate Std. Error t value Pr(>|t|)
## (Intercept)    -0.1397     0.1991  -0.701  0.48326
## X               1.1286     0.3572   3.159  0.00163 **
## A               3.9496     0.1954  20.216  < 2e-16 ***
## ---
## Signif. codes: 0 '***' 0.001 '**' 0.01 '*' 0.05 '.' 0.1 ' ' 1
##
## Residual standard error: 6.295 on 997 degrees of freedom
## Multiple R-Squared: 0.473, Adjusted R-squared:  0.472
## Wald test: 224 on 2 and 997 DF, p-value: < 2.2e-16
```

Now the standard error is 4.41× larger than the one estimated by conditioning on U. Controlling for the competing exposure variable A in the model resulted in a standard error that is 0.85× as large as the standard error in the analysis that did not include A. This represents a substantial gain in precision.

Including the confounder U in the model, in addition to A, would make an even larger contribution to the precision of the estimate. Obviously this would be impossible given that U is supposed to represent an unobserveable confounder, but it is included in the simulated data set. For the sake of illustration, the ivreg() results are given below.

```
##
## Call:
## ivreg(formula = Y ~ X + A + U | I + A + U, data = data)
```

```
##
## Residuals:
##       Min       1Q    Median        3Q       Max
## -14.95969  -2.64426   0.02956   2.55427  10.53529
##
## Coefficients:
##              Estimate Std. Error t value Pr(>|t|)
## (Intercept)  -0.07984    0.12515  -0.638    0.524
## X             1.08175    0.22663   4.773 2.08e-06 ***
## A             3.71698    0.12162  30.562  < 2e-16 ***
## U             4.78224    0.26671  17.930  < 2e-16 ***
## ---
## Signif. codes:  0 '***' 0.001 '**' 0.01 '*' 0.05 '.' 0.1 ' ' 1
##
## Residual standard error: 3.953 on 996 degrees of freedom
## Multiple R-Squared: 0.7924, Adjusted R-squared: 0.7918
## Wald test: 1175 on 3 and 996 DF, p-value: < 2.2e-16
```

With both A and U included in the model, the standard error is 2.8× larger than the one estimated by conditioning on U, and 0.54× as large as the standard error from the 2SLS analysis that did not include A or U. Even though IV methods are designed to protect against unobserved confounding, if researchers are able to identify and measure known or suspected confounders, they should do that and include them in the analysis. Doing so can result in a large improvement in precision.

Why use IVA at all if one has measures of *all* the suspected confounding variables? If conditional ignorability can be safely assumed, then this method would indeed be a poor choice due to its low efficiency. To use IV in such a circumstance is a waste of valuable data; one can achieve much more precise estimates given a particular sample size using conditioning-based methods. But IV is warranted when the researchers are not comfortable with the conditional ignorability assumption or when there are concerns about the validity or precision of measurement of those confounders, especially when the sample size is large enough that the efficiency deficit is less bothersome or when a strong instrument can be found.

One note of caution about incorporating auxiliary variables into the IVA is shown by the DAG in Figure 6.8. Here, variable C is a collider with both the instrument and the response variable as causes. If the analyst mistakenly believes that C is a competing exposure or potential confounder and incorporates it into the IV model to increase efficiency, a back-door path will be opened from between I and Y. In this case, I will no longer be an instrument, and biased treatment effect will result.

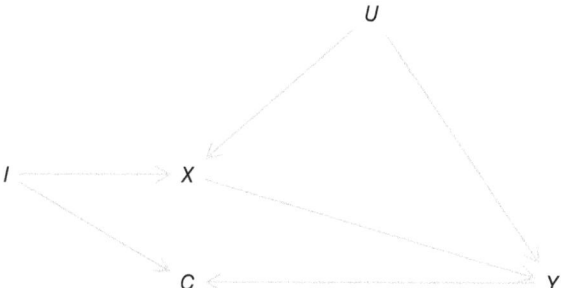

Figure 6.8 Adding auxiliary variables to IV models risks creating collider bias if the variable role is misidentified. Conditioning on variable *C* opens a back-door path between *I* and *Y*, introducing bias into the treatment effect estimate

Note. IV = instrumental variable.

Sample size issues

The precision and statistical power achieved by IVA depends on the strength of the instrument, the sample size and the impact of any other covariates included in the stage 2 model. Boef et al. (2014) examined the bias–variance trade-off offered by IV methods: bias reduction, yes, but potentially at the cost of a vast amount of precision. There are times when a biased but low-variance estimate will be more accurate, on average, than a high-variance but unbiased estimate, and therefore would be preferred (James et al., 2013). Boef et al. (2014) derived a method for determining the minimum sample sizes in which IV methods are preferred over ordinary regression models, as determined by two factors: (1) the degree of confounding bias and (2) the strength of the instrument. In the weak confounding/weak instrument condition, a sample size of more than $n = 10,000$ was required before IV methods became favourable. Even the severe confounding/strong instrument condition exhibited a threshold sample size of $n = 2,600$ before IV methods were preferred. IVA should therefore be regarded as a large sample technique. However, their analysis did not account for the beneficial effect of including auxiliary covariates in the stage 2 model, which can substantially reduce the required sample size.

Measurement error

This section will examine how imperfect measurement affects IVA. Neither measurement error in the exposure variable nor in the instrument create bias, though they do reduce efficiency.

Imperfect measurement of the exposure

Measurement error in the exposure variable is an additional source of endoge-neity beyond classical confounding that IV methods can address (Bollen, 2012). As presented in Chapter 2, measurement error bias is fundamentally equivalent to confounding bias because the unobservable true score is a confounder of the relationship between the observed score and the response variable. The DAG in Figure 6.9 illustrates the situation that occurs when the $X \to Y$ relationship is disturbed by an unobserved confounder and when X is measured with error. And of course, perfect measurement is an unattainable ideal, so X will always be measured with some error in reality.

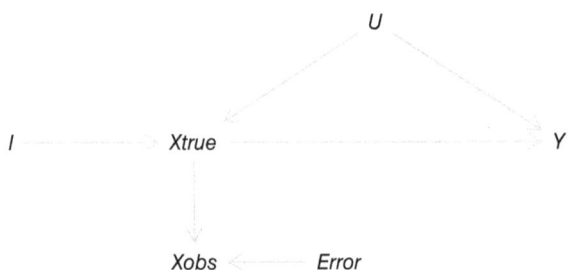

Figure 6.9 Example of a DAG in which the exposure is measured with error. *Xtrue* is the exposure true score, *Xobs* is the exposure observed score, *Error* is the measurement error, *Y* is the response, *U* is an unobservable confounder and *I* is the instrument

Note. DAG = directed acyclic graph.

To understand how IVA removes this type of bias, it is helpful to separately con-sider the two stages of the estimation process and their associated DAGs. The DAG for the first stage is shown in Figure 6.10. The stage 1 model seeks to estimate the relationship between the instrument (*I*) and the observed X variable so that pre-dicted values can be saved for use in the stage 2 model. From the perspective of

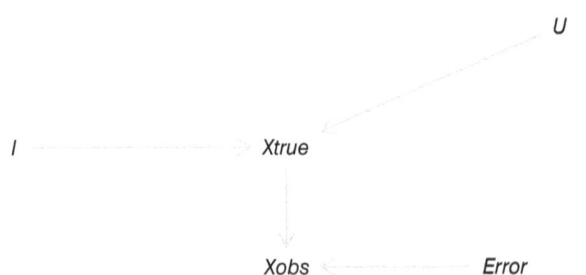

Figure 6.10 DAG for stage 1 of the IV estimator when X is measured with error

Note. DAG = directed acyclic graph; IV = instrumental variable.

this analysis, *I* is the exposure and *Xobs* is the response variable. Fitting the stage 1 regression model to the data isolates the total effect of *I* on *Xobs*. This path happens to be mediated (by the unobserved *Xtrue*), but this is of no relevance because conditioning on mediators is unnecessary. As discussed in Chapter 3, one should not condition on mediators when the total effect is of interest. The predicted values of *Xobs* generated from this model will be purged of influence from *U* and *Error*. The whole purpose of the stage 1 model is to remove the component of *X* that is confounded by *U*. This purpose is accomplished even when *X* is measured imperfectly (Bollen, 2012).

The stage 2 DAG is shown in Figure 6.11. The predictions generated from the stage 1 model, *Xpredicted*, are a function of the instrument and the exposure variable's unobservable true score, which completely mediates the *I* → *Xpredicted* path. These predictions have also been purged of the measurement error component of *Xobs* because the errors are independent of *I*. The new exposure variable in the stage 2 model is *Xpredicted*, and it is clear from the diagram that obtaining an unbiased estimate of its causal effect on *Y* is straightforward since there is no confounding of the *Xpredicted* → *Y* path.

Figure 6.11 DAG for stage 2 of the IV estimator when *X* is measured with error

Note. DAG = directed acyclic graph; IV = instrumental variable.

The bias caused by measurement error in *X* is therefore removed by IV methods. The imprecise measurement of *X* does weaken *I*'s strength as an instrument, which manifests as some precision loss in estimating the causal effect of *X* on *Y*. There is no free lunch in statistics; measurement error always comes at a cost. But lunch is nearly free in this case because the amount of lost precision is quite minimal.

I added a new variable, *Xobs*, to the simulated data set with a reliability coefficient of 0.5. The poor measurement of *X* makes *I* a weaker instrument. The R^2 statistic in the stage 1 model changed from 0.054 to 0.031. The results of the 2SLS analysis using the ivreg() function are given below.

```
##
## Call:
## ivreg(formula = Y ~ Xobs | I, data = data)
```

```
##
## Residuals:
##       Min       1Q   Median       3Q      Max
## -26.4193  -5.1226   0.1138   5.4229  26.9477
##
## Coefficients:
##                Estimate Std. Error t value Pr(>|t|)
## (Intercept)    -0.0805     0.2563  -0.314  0.75355
## Xobs            1.2145     0.4305   2.821  0.00488 **
## ---
## Signif. codes: 0 '***' 0.001 '**' 0.01 '*' 0.05 '.' 0.1 ' ' 1
##
## Residual standard error: 8.104 on 998 degrees of freedom
## Multiple R-Squared: 0.1258, Adjusted R-squared: 0.1249
## Wald test: 7.96 on 1 and 998 DF, p-value: 0.004877
```

Compared to an IV model where X is measured perfectly, the standard error of this model is only 1.03× larger.

IV methods could be used to correct for measurement error in X even in the absence of other confounding, but this is generally not recommended because the precision loss typically results in higher MSE than the measurement error bias itself would cause unless the instrument is exceptionally strong. In this case, the cure is worse than the disease. Researchers should instead obtain multiple measurements of **X** and use structural equation modelling techniques to address the measurement error (Bollen, 2014).

Measurement error in the instrument

Figure 6.12 is a DAG in which the instrument is measured with error. The instrument's true score, *Itrue*, is now a confounder of the *Iobs* → X path. This implies that the stage 1 model will produce a biased estimate of the causal effect of the instrument on the exposure and also that the predicted values of X derived from that model will be biased.

This is true, but irrelevant to the actual goal of achieving an unbiased estimate of the causal effect of X on Y, which can be accomplished regardless of whether the stage 1 model produces a biased estimate. The only requirement is that the first model's predicted values must be free of influence from any confounders of the $X → Y$ path. And this will be the case as long as I is actually an instrument. The error-contaminated version of I still fulfils the requirements for IVA: it is correlated with the exposure and has no influence on the response except through the exposure. In this case, the correlation between the observed instrument and X is spurious (due to the *Iobs* ← *Itrue* → X confounding path), but this is of no consequence.

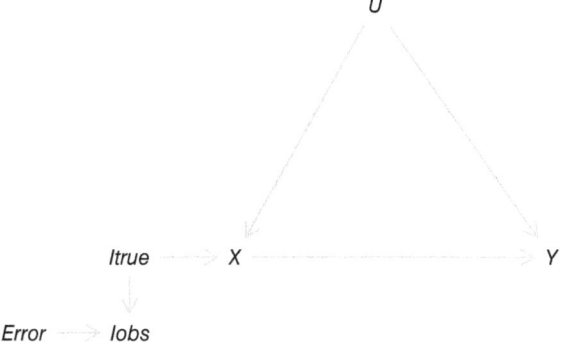

Figure 6.12 Example of a DAG in which the instrument is measured with error. *X* is the exposure, *Y* the response, *U* an unobservable confounder, *Itrue* the instrument true score, *Iobs* the instrument observed score and *Error* the measurement error

Note. DAG = directed acyclic graph.

I appended variable *Iobs* to the simulated data set, which is a poorly measured version of *I* with a reliability coefficient of 0.5. The results of the 2SLS analysis using this variable as the instrument are given below.

```
##
## Call:
## ivreg(formula = Y ~ X | Iobs, data = data)
##
## Residuals:
##      Min       1Q   Median       3Q      Max
## -24.1009  -4.5520   0.2038   4.7458  23.0660
##
## Coefficients:
##              Estimate Std. Error t value Pr(>|t|)
## (Intercept) -0.08816    0.22820  -0.386  0.69933
## X            1.76824    0.51014   3.466  0.00055 ***
## ---
## Signif. codes: 0 '***' 0.001 '**' 0.01 '*' 0.05 '.' 0.1 ' ' 1
##
## Residual standard error: 7.213 on 998 degrees of freedom
## Multiple R-Squared: 0.3074, Adjusted R-squared: 0.3067
## Wald test: 12.01 on 1 and 998 DF, p-value: 0.0005505
```

The standard error from this model in which the instrument is poorly measured (reliability = 0.5) is 0.510, which is 1.22× larger than the one in the model with no measurement error in *I*. Measurement error in the instrument typically imposes a much higher penalty than poor measurement of the exposure on estimation precision of the causal effect.

Local average treatment effects

In contrast with conditioning methods, which can estimate the ATE (or subgroup-specific effects such as the ATT), IV methods estimate the treatment effect specifically for the subpopulation of subjects whose treatment assignment is responsive to the instrument. This group is often called the *compliers* (Greenland, 2000). Referring back to the example of cigarette taxes as an instrument for smoking, some participants will smoke regardless of how expensive cigarette taxes become. These are called *always-takers* because their probability of being treated is 100%. There are also *never-takers*, who would not smoke even if the cigarette tax rate was zero. And one must admit, as an abstract logical possibility, the potential existence of *defiers*, people whose probability of treatment opposes the instrument (Barnard et al., 2003). These individuals would be more likely to smoke when cigarette taxes are high than they would when taxes are low. Defiers would create a non-removable bias, and so the assumption of a monotonic relationship between the instrument and the exposure variable is required in order for IV methods to yield consistent estimates. This assumption that the relationship between the instrument and the exposure variable must be unidirectional or zero for all subjects is called the monotonicity assumption. The monotonicity assumption prohibits the existence of defiers.

It is likely that compliers, always-takers, never-takers and defiers differ from one another in many ways. If these differences extend to their potential outcomes (and they probably do), there may be considerable treatment effect heterogeneity across these categories. The causal effect of smoking on lung function could be different for compliers of the cigarette tax instrument than it might be for always-takers or never-takers. For example, perhaps always-takers are wealthier and receive higher quality and more regular preventative medical care than compliers, who are by definition more cost-sensitive in their behaviour. Alternatively, it is possible that always-takers are more strongly addicted, perhaps because they prefer flavoured products or conceivably because they have a higher predisposition towards addictive behaviours. Maybe never-takers have higher rates of pre-existing lung disease, such as asthma, and cannot 'afford' to further jeopardise their breathing with tobacco smoke. Or maybe never-takers are more active in recreational sports and fitness activities and, as a consequence, would suffer less health impairment from smoking. The point is that the magnitude of the causal effect itself could vary across these categories.

Chapter 4 defined the ATE as the marginal treatment effect for the entire population. The ATE cannot be estimated by IV methods unless it is either assumed that there is no treatment effect heterogeneity or that compliance is ignorable (i.e. unrelated to potential outcomes). Both of these are strong and generally implausible assumptions that should not be relied upon. The treatment effect that can be credibly produced via IV methods is a local average treatment effect **(LATE)**, which applies specifically to the compliers and cannot be generalised beyond that group. For this reason, IV methods are deemed to have strong internal validity but relatively weak external validity. It is often unclear to what subpopulation, exactly, the resulting treatment effect estimate can be generalised (Greenland, 2000).

It is also important to note that compliance status is always related to the specific instrument and how it influences behaviour. The subpopulation of compliers for a cigarette tax instrument will be different from those who would comply with a 'distance to the nearest convenience store' instrument, which could be different from those who would comply with a 'smoking ban in public places' instrument. In other words, compliance is not a fixed property of a person but rather of the interaction between the person and the inducement offered by a specific instrument. As a result, studies of the same phenomenon or treatment utilising different instruments could be estimating meaningfully and systematically different treatment effects.

Assumptions of instrumental variable analysis

The assumptions of IVA are as follows:

- *Relevance:* The instrument must be related to the exposure variable. The magnitude of this relationship is known as the *strength* of the instrument.
- The *exclusion restriction:* The instrument only affects the outcome through the exposure variable. Its effect must be fully mediated by the exposure.
- *Exchangeability* (no confounding): The effect of the instrument on the outcome must not be confounded.
- *Monotonicity* (no defiers): The effect of the instrument on the exposure variable cannot be bidirectional. For example, an instrument that induces some people to smoke cigarettes cannot induce other people to quit.

Further Reading

Angrist, J. D., Imbens, G. W., & Rubin, D. B. (1996). Identification of causal effects using instrumental variables. *Journal of the American Statistical Association*, *91*(434), 444–455. https://doi.org/10.2307/2291629

This article considers the application of instrumental variables analysis to a 'broken' randomised experiment and shows how IVs fit into Rubin's causal model.

Angrist, J. D., & Krueger, A. B. (2001). Instrumental variables and the search for identification: From supply and demand to natural experiments. *Journal of Economic Perspectives, 15*(4), 69–85. https://doi.org/10.1257/jep.15.4.69

This exceptionally well-written article discusses the history of instrumental variables analysis and its application to the study of natural experiments. The article contains a bibliography with references to over a thousand papers employing instrumental variables methods.

Bennett, D. A. (2010). An introduction to instrumental variables analysis: Part 1. *Neuroepidemiology, 35*(2), 237–240. https://doi.org/10.1159/000319455

This article is a brief introduction to instrumental variables analysis for medical researchers. The paper begins with a discussion of the general problem of confounding and discusses how instrumental variables can provide a viable alternative to experimental designs.

Greenland, S. (2000). An introduction to instrumental variables for epidemiologists. *International Journal of Epidemiology, 29*(4), 722–729. https://doi.org/10.1093/ije/29.4.722

This paper introduces the instrumental variables analysis to epidemiological researchers, focusing on its application to broken experiments. A non-parametric version of instrumental variables analysis now known as complier average causal effects (CACE) is described.

Heckman, J. (1997). Instrumental variables: A study of implicit behavioral assumptions used in making program evaluations. *Journal of Human Resources, 32*(3), 441–462. https://doi.org/10.2307/146178

This paper critiques instrumental variable analysis, arguing that the validity of conclusions drawn from these methods depends on strong assumptions about how people process information. The paper argues that instrumental variables methods do not estimate relevant values when subjects are able to anticipate their benefit from treatment.

7

REGRESSION DISCONTINUITY DESIGN

Chapter Overview

This chapter introduces the regression discontinuity design (RDD) for causal inference (Bloom, 2012). The regression discontinuity approach has much more in common with randomised experiments than with observational studies. In both, treatment is assigned by a process that is fully understood, though in RDD this process is at least partially deterministic rather than random. Subjects must be allocated to treatment conditions based on their scores on a pretreatment variable. The rule must be quantitative in nature and consistently applied. This implies that the rule must not be adjusted in order to provide treatment to participants who are perceived to be especially deserving or in need. In the strong version of RDD, called sharp RDD, the rule must be strictly enforced with no exceptions. But causal inference is still possible from an alternative version known as fuzzy RDD, in which the assignment rule is probabilistic rather than deterministic, and there may be exceptions to the assignment rule (Venkataramani et al., 2016). Some example scenarios in which RDD could be applied are as follows:

1   Health insurance companies often require enrolment in mandatory health coaching for clients with health wellness scores below some threshold value. The company would like to know how much health coaching reduces claims.
2   Food assistance programmes (e.g. Women, Infants and Children [WIC]) are made available to people who have a sufficiently low adjusted income. Nutritionists would like to know if WIC reduces the incidence of low birthweight and failure to thrive in early childhood.
3   Students with sufficiently high cognitive ability are assigned to a gifted and talented education programme. School district administrators want to understand the gifted programme's impact on academic achievement.

The techniques presented in the previous chapters have been based on the context of observational studies, where the researcher does not control treatment assignment. Research subjects may select themselves into treatment conditions (e.g. smoking), or are perhaps 'assigned' on the basis of factors beyond their control (e.g. unemployment). Treatment assignment is a *fait accompli* when the data become available to the analyst. Reaching a valid causal inference reduces to a game of reconstituting the relevant facts of the assignment mechanism. This is seen most clearly in PSA, where the propensity scores are, quite literally, a model of how background variables manifest as treatment assignment.

Contrast this with lab experiments or randomised controlled trials. Here the assignment mechanism is known. Causal inference is a feature of the design of such a study rather than the conditioning strategy (Campbell & Stanley, 1966). Randomisation can be understood from many perspectives: it makes impossible any back-door paths connecting treatment assignment and the outcome; it creates ignorablility by rendering the potential outcomes independent of treatment assignment; it serves as the strongest possible instrument. Randomisation produces equivalence in expectation (and all higher order moments too) on all pretreatment background variables, including potential outcomes.

Causal inference from the RDD may initially seem impossible. For randomised experiments, the utility of equivalence in expectation on all pretreatment variables is intuitively obvious. The fact that treatment and control subjects are identical (in expectation) on *everything* prior to treatment makes the attribution of any post-treatment differences to the treatment itself quite natural. After all, the only systematic difference between the groups is their exposure to treatment. There is nothing else to blame. The logical coherence of the inference follows directly from the understanding that randomisation-derived equivalence rules out a huge set of potential, non-causal explanations for observed effects (Campbell & Stanley, 1966). Such is not the case for the RDD. In RDDs, the treatment and the control subjects are *not* equivalent. They differ systematically on the forcing variable by design, and this almost certainly implies non-equivalence on many other background variables as well, some of which are likely to be confounders. The RDD is, in some ways, the opposite of a randomised experiment.

Consider the example of the gifted programme evaluation. A naive estimate of the treatment effect would involve a simple comparison of post-treatment academic achievement outcomes for the treated (e.g. gifted programme) versus untreated (regular classroom) students. But this comparison would yield no real information and is, in fact, tautological. If the academic ability (F) has any relationship at all with achievement (Y), then, of course, treated students will tend to have higher achievement regardless of whether the treatment is effective or not.

RDD allows for the causal effect of treatment to be estimated in spite of this non-equivalence. Treatment assignment must be allocated on the basis of a thresholding process in which a particular cut score on a continuous variable determines whether each subject is assigned to treatment or control. This design does not exploit natural variability or self-selection into treatment conditions. It exploits the known features of the assignment mechanism, while allowing some degree of self-selection or non-compliance under the fuzzy RDD model.

The forcing variable and treatment assignment

The variable that determines treatment assignment is called the **forcing variable** (F) or *running variable*. Let z be some threshold value. In sharp RDD, all cases with $F < z$ are assigned to one treatment status (D), while cases with $F \geq z$ are assigned to the other treatment status. In fuzzy RDD, treatment assignment is a probabilistic function of the forcing variable such that there is a mixture of treatment and control cases on either side of the boundary. Both the sharp and fuzzy variants of RDD will be discussed in this chapter and illustrated with figures.

The RDD is a natural way to evaluate the efficacy of social and educational poli-cies. Treatment can be allocated to those most in need, unlike randomised designs in which some needy subjects are assigned to the control group and thereby deprived of a potentially beneficial treatment. From an ethical standpoint, the RDD can be far more defensible than randomisation in many contexts. Implementing RDD as an evaluation model can be less disruptive to the interventionist goals of a programme or policy. Further, its assumptions are quite mild compared to methods based on achieving conditional ignorability through conditioning.

Another strength of the RDD is that it is largely a graphical method, which not only facilitates communication of the results but also allows for an easy appraisal of some of the method's assumptions. Unlike some procedures which bury many of the details in multivariable mathematical abstraction, the RDD is relatively transparent and intuitive. In many cases, the treatment effect is readily apparent upon visual inspection, lending additional credibility to the findings.

In spite of these strengths, RDD has historically been an underutilised design. It was first proposed by Thistlethwaite and Campbell (1960) and was described in Campbell and Stanley's (1966) enormously popular text on causal inference. The virtues of the design are now better understood due to the formal mathematical basis that has been bestowed on it via the potential outcomes framework. As a result, it is experiencing something of a renaissance in many fields (de la Cuesta & Imai, 2016).

Sharp RDD

Figure 7.1 displays a DAG or causal diagram describing the sharp RDD. Note that the forcing variable (F) is the only confounder of the treatment (D) to response (Y) path. There are no other influences on D besides F. This is one of the key assumptions of sharp RDD, and it is trivially easy to check. This is because any influence on D besides F would create exceptions to the assignment rule – a situation that requires the fuzzy version of RDD.

In contrast, the forcing variable F and the response variable Y can be subject to confounding by any number of observable or unobservable confounders. By assump-tion, the only paths by which these potential confounders can affect Y is through F. This is because, in sharp RDD, D is *only* caused by a thresholding process applied to F. This means that conditioning on the forcing variable is sufficient for blocking all potential back-door paths between D and Y, including the $F \rightarrow D$ direct effect. From this perspective, sharp RDD is just another straightforward conditioning problem. The key assumption is that the $D \rightarrow Y$ path is unconfounded by anything other than F. (Note that this assumption is relaxed in fuzzy RDD.)

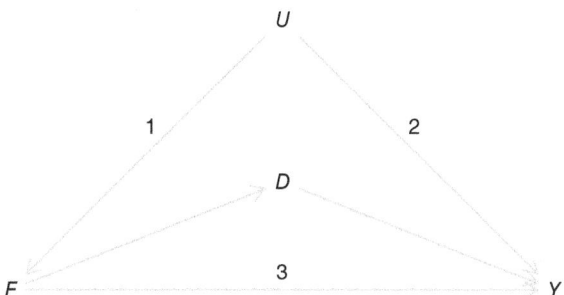

Figure 7.1 Example of a DAG for a sharp regression discontinuity design. F = Forcing variable; D = Treatment status assigned by thresholding the forcing variable; Y = Response; U = Unobserved confounder. Paths 1, 2 and 3 are typical but not required

Note. DAG = directed acyclic graph.

The causal effect of D on Y is identifiable by conditioning on F. The basic intuition of the RDD is that treatment assignment is ignorable at the boundary (z). At the boundary, the difference between the treatment and control units on F, and therefore on all other potential confounders, is zero. This means that the treatment effect can be estimated by simply comparing the expected value of Y for treated subjects against the expected value of Y for the control subjects at the boundary. But there is a challenge: there is complete lack of common support. If treatment is assigned to cases with $F > z$, then there will be no treated cases with $F > z$ and no control cases with $F > z$. This implies that the non-parametric conditioning methods that were examined in Chapter 2, such as matching, weighting and subclassification, cannot be used and that parametric conditioning via a statistical model is hazardous due to the required extrapolation (Hahn, 1977). The RDD can be understood from the perspective of Rubin's causal model (Bloom, 2012). The key assumption for both sharp and fuzzy RDD is that the potential outcomes are continuous near the boundary. This implies that the potential outcomes do not experience the discontinuity – only treatment assignment does, and that any difference in potential outcomes across levels of treatment become arbitrarily small as the boundary is approached from either side, as shown by Equation (7.1) (Hahn et al., 2001).

$$\lim_{F \to z+} (Y_0, Y_1) = \lim_{F \to z-} (Y_0, Y_1) \tag{7.1}$$

Figure 7.2 displays three different data conditions for the strict RDD. In the left panel, all the control cases lie just below the boundary and all the treated cases just above it. In this situation, the treatment effect could be computed as a straightforward comparison of treatment and control means using all the data, because nature has already conditioned on F. This would be a highly unusual situation to observe in real life. In the middle panel, a small fraction of the data lie near the boundary, but most are further away. And in the right panel, there are hardly any data near the boundary. Extrapolation from the mass of the data to the boundary becomes increasingly necessary as the fraction of data near the boundary decreases. In middle panel, the treatment effect could be estimated using the small fraction of the data with $F \approx z$, but this would waste a large fraction of the data and result in a loss of precision. In the right panel, extrapolation is an absolute requirement. There are only two cases near the boundary.

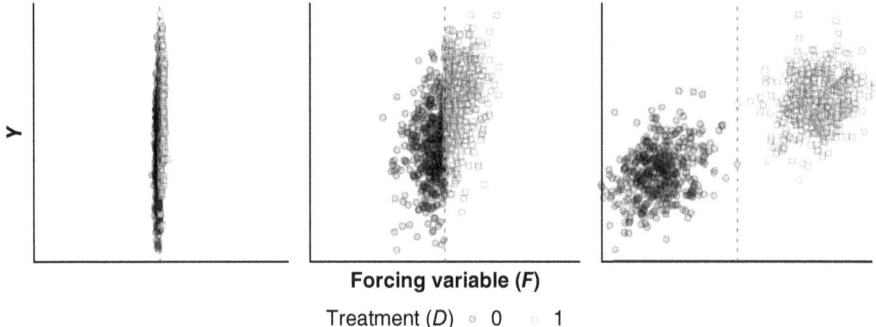

Forcing variable (F)

Treatment (D) ○ 0 ▫ 1

Figure 7.2 Necessity of extrapolation to estimate the magnitude of the discontinuity at the boundary. Left panel: No extrapolation required because the entire sample lies near the boundary. Middle panel: Some data near the boundary. Right panel: Almost no data near the boundary

Some type of model-based projection or extrapolation is required for non-boundary cases to contribute information to the estimation process. Analysts typically rely on linear regression (often incorporating non-linear components) or a non-parametric regression smoothing approach to estimate the expected values of Y at the boundary for the two groups (Imbens & Lemieux, 2008). Parametric regression weights all the cases equally no matter how distant they may be from the cut point, whereas non-parametric smoothing differentially weights the cases such that cases lying near the cut point carry more influence than those located far away from it. Both approaches assume that the relationship between F and Y, which is well-identified in regions with adequate data, can be extrapolated to the boundary.

In either case, a contrast of the model-predicted values of Y at the boundary for the treatment and control groups estimates the treatment effect (Hahn et al., 2001).

Extrapolation via parametric regression

Figure 7.3 displays an example scatter plot of data from an RDD with a superimposed regression fit. The regression line is used to project from the data to the boundary, allowing the predicted values of Y at the boundary under the treatment and control conditions to be estimated. In this case, the projection is successful because the underlying functional form is linear. The regression model is correctly specified and is therefore able to identify the causal effect, which is visually identified as the vertical magnitude of the regression line discontinuity at the boundary.

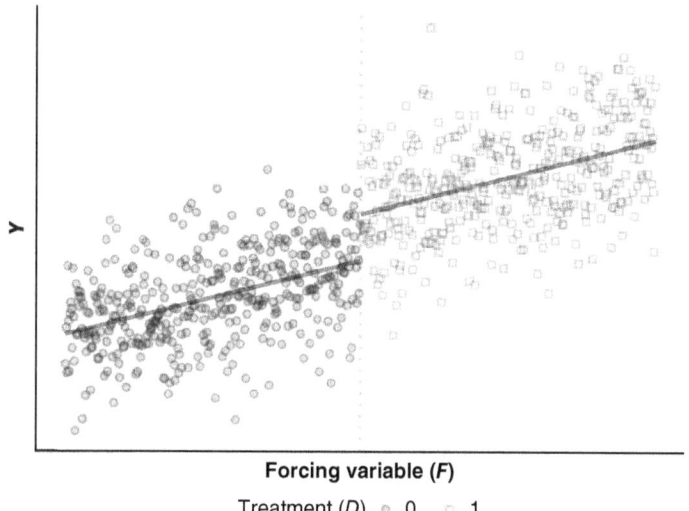

Forcing variable (F)

Treatment (D) ○ 0 ○ 1

Figure 7.3 Illustration of a strict regression discontinuity design. x-axis: The forcing variable (F). y-axis: The response (Y). Point shape and colour: Treatment status (D). All cases below the threshold on F are assigned to control, cases above it are assigned to treatment. The regression line indicates the relationship between F and Y. The treatment effect is represented by the discontinuity of the regression at the threshold value

The simplest regression model for performing linear extrapolation is shown in Equation (7.2). Parameter b_2 is a main effect for the binary treatment variable D, which shifts the regression line vertically and creates the discontinuity at the treatment status boundary. If D is coded {0,1} for control and treatment, respectively, then the inclusion of b_2 implies an intercept of b_0 for control subjects and $b_0 + b_2$ for treatment subjects.

$$Y_i = b_0 + b_1(F_i) + b_2(D_i) + e_i \qquad (7.2)$$

Fitting this model to the data produces the following output:

```
## 
## Call:
## lm(formula = Y ~ F + D, data = data)
```

```
##
## Residuals:
##       Min        1Q   Median       3Q       Max
## -20.6477   -4.7324  -0.3897   4.9251   27.1080
##
## Coefficients:
##               Estimate Std. Error t value Pr(>|t|)
## (Intercept)     0.1545     0.5681   0.272    0.786
## F               2.9440     0.2198  13.396  < 2e-16 ***
## D               7.4624     1.0163   7.343 5.18e-13 ***
## ---
## Signif. codes: 0 '***' 0.001 '**' 0.01 '*' 0.05 '.' 0.1 ' ' 1
##
## Residual standard error: 7.187 on 797 degrees of freedom
## Multiple R-squared: 0.6696, Adjusted R-squared: 0.6687
## F-statistic: 807.5 on 2 and 797 DF, p-value: < 2.2e-16
```

The estimated effect of D is an estimate of the causal effect of treatment, $b_2 = 7.462$ (1.016), 95% CI [5.470, 9.454], $p < .001$. The true causal effect in this simulated data was 8.00, which is well within the confidence interval. The ability of the simple model in equation (7.2) to produce unbiased and consistent estimates of the causal effect is based on two additional technical assumptions. Unfortunately, the plausibility of these assumptions in many situations is questionable. The first is the assumption that the functional form relating the response to the forcing variable is linear. Figure 7.4 illustrates the possible consequences of violating this assumption.

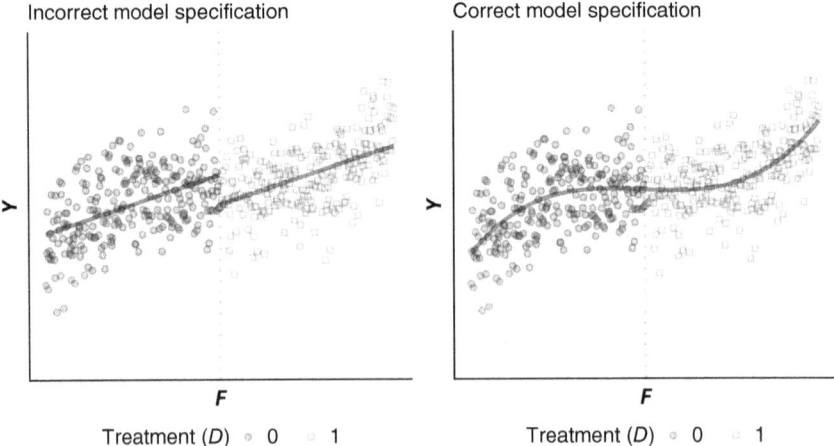

Figure 7.4 Misspecification of the functional form relating the response to the forcing variable can lead to a biased estimate of the treatment effect. Left panel: An incorrect assumption of linearity leads to a large apparent discontinuity. Right panel: The correct non-linear functional form reveals the absence of a discontinuity

Non-linearity can create an apparent discontinuity (and therefore, causal effect of treatment) when none actually exists. The left panel of the figure seems to indicate a large treatment effect as a consequence of imposing a linear model on non-linear data. The right panel shows how the true treatment effect of zero is recovered when the proper non-linear model is fit to the data.

Certain classes of non-linear relationships can be modelled using linear regression by including polynomial terms in the model. In this case, the correct model includes quadratic (squared) and cubic (third-power) terms, with the form $Y_i = b_0 + b_1(F_i) + b_2(F_i^2) + b_3(F_i^3) + b_4(D_i) + e_i$. The lm() output from fitting this model to the data is given below.

```
##
## Call:
## lm(formula = Y ~ F + I(F^2) + I(F^3) + D, data = data)
##
## Residuals:
##      Min      1Q   Median      3Q      Max
## -28.7803  -6.9854  -0.3907  7.1928  27.9958
##
## Coefficients:
##               Estimate Std. Error t value Pr(>|t|)
## (Intercept) 80.37682    1.40837  57.071  < 2e-16 ***
## F           -0.57681    0.99463  -0.580    0.562
## I(F^2)       0.05285    0.09611   0.550    0.583
## I(F^3)       0.39289    0.06305   6.231 9.91e-10 ***
## D           -0.31195    2.45501  -0.127    0.899
## ---
## Signif. codes: 0 '***' 0.001 '**' 0.01 '*' 0.05 '.' 0.1 ' ' 1
##
## Residual standard error: 10.29 on 495 degrees of freedom
## Multiple R-squared: 0.3943, Adjusted R-squared: 0.3894
## F-statistic: 80.56 on 4 and 495 DF, p-value: < 2.2e-16
```

The estimated treatment effect of D is b_4 = –0.312 (2.455), 95% CI [–5.124, 4.500], p = .899. The true causal effect of zero was captured in the confidence interval.

Another technical assumption is that the functional form relating Y to F is the same for both levels of treatment. This could be termed the 'no interaction' assumption (Trochim et al., 1991). Figure 7.5 shows how the violation of this assumption can lead to a biased estimate of the treatment effect by suggesting a discontinuity where none exists.

The 'no interaction' assumption can be relaxed by adding interaction terms between F and D to the regression model. For example, fitting a model of the form $Y_i = b_0 + b_1(F_i) + b_2(D_i) + b_3(F_i \times D_i) + e_i$ to the data yields the following result:

```
##
## Call:
## lm(formula = Y ~ F * D, data = data2)
##
## Residuals:
##     Min      1Q  Median      3Q     Max
## -5.7306 -1.4651 -0.0457  1.4959  5.4399
##
## Coefficients:
##             Estimate Std. Error t value Pr(>|t|)
## (Intercept)   9.8613     0.4587  21.499   <2e-16 ***
## F             4.9275     0.1732  28.446   <2e-16 ***
## D            -4.5381     0.4889  -9.283   <2e-16 ***
## F:D          -5.0278     0.1912 -26.292   <2e-16 ***
## ---
## Signif. codes: 0 '***' 0.001 '**' 0.01 '*' 0.05 '.' 0.1 ' ' 1
##
## Residual standard error: 2.066 on 496 degrees of freedom
## Multiple R-squared: 0.8294, Adjusted R-squared: 0.8284
## F-statistic: 804 on 3 and 496 DF, p-value: < 2.2e-16
```

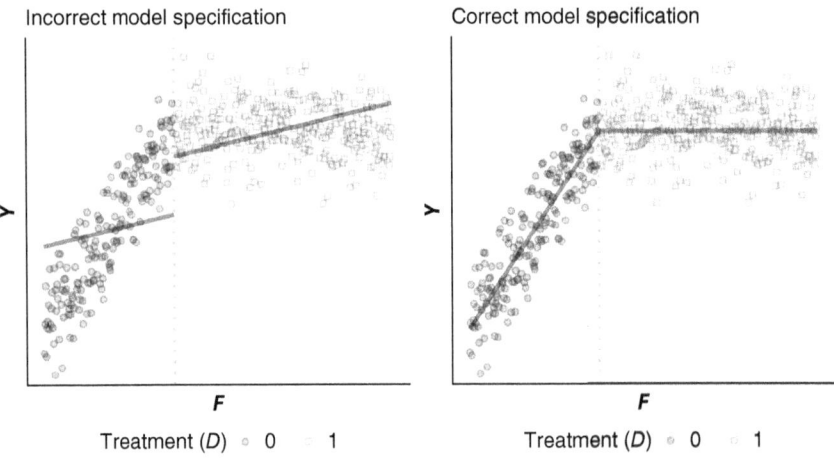

Figure 7.5 Violating the 'no interaction' assumption can lead to a biased estimate of the treatment effect. Left panel: An incorrect assumption of a common slope of Y on F across levels of treatment leads to a large apparent discontinuity. Right panel: The correct model reveals the absence of a discontinuity

One must be very careful to interpret the results from this model properly. Because of the presence of the interaction parameter, the coefficient b_2 no longer estimates the treatment effect. The treatment effect is defined specifically at the boundary, z, but b_2 estimates the difference between the expected values of Y for the treatment and control groups at $F = 0$. The interaction term implies that this difference in expected values will vary across F.

Estimating the treatment effect at the boundary requires centering F at z. I created a new variable called Fz by subtracting z from all the values of F. Fitting the model $Y_i = b_0 + b_1(F_i - z) + b_2(D_i) + b_3[(F_i - z) \times D_i] + e_i$ to the data produces the following output:

```
##
## Call:
## lm(formula = Y ~ Fz * D, data = data2)
##
## Residuals:
##     Min      1Q  Median      3Q     Max
## -5.7306 -1.4651 -0.0457  1.4959  5.4399
##
## Coefficients:
##                Estimate Std. Error t value Pr(>|t|)
## (Intercept)     4.9338     0.3005  16.417   <2e-16 ***
## Fz              4.9275     0.1732  28.446   <2e-16 ***
## D               0.4898     0.3811   1.285    0.199
## Fz:D           -5.0278     0.1912 -26.292   <2e-16 ***
## ---
## Signif. codes:  0 '***' 0.001 '**' 0.01 '*' 0.05 '.' 0.1 ' ' 1
##
## Residual standard error: 2.066 on 496 degrees of freedom
## Multiple R-squared: 0.8294, Adjusted R-squared: 0.8284
## F-statistic: 804 on 3 and 496 DF, p-value: < 2.2e-16
```

The parameter b_2 properly estimates the treatment effect after centring, $b_2 = 0.49$ (0.381), 95% CI [–0.257, 1.237], $p = .199$. The true value of zero was captured in the interval.

It is possible to relax both the linearity and the 'no interaction' assumption simultaneously by including a set of polynomial terms as well as their interactions with treatment in the model. However, the functional form must be correctly specified for the model-predicted values of Y at the boundary to be accurate. Such a model would be given by Equation (7.3).

$$Y_i = b_0 + b_1(F_i) + b_2(F_i^2) + b_3(F_i^3) + b_4(F_i \times D_i) + b_5(F_i^2 \times D_i) + b_6(F_i^3 \times D_i) + b_7(D_i) + e_i$$

(7.3)

In this model, parameters b_1 through b_3 describe the non-linear functional form of the relationship between F and Y for the control group, interaction parameters b_4 through b_6 allow this form to deviate for the treatment group and b_7 estimates the treatment effect (assuming that F has been centred prior to analysis to the threshold z). Of course, the inclusion of more polynomial terms is possible. However, Gelman and Imbens (2019) critiqued the use of higher order polynomials (e.g. third order and higher) for RDD estimation, arguing that they often overfit the data and result in estimates that are noisy and highly sensitive to the order of the polynomial function. They recommended that analysts instead use non-parametric smoothing to calculate the treatment effect.

Non-parametric approach

An alternative estimation strategy is to employ a non-parametric approach for calculating the predicted values of Y at the boundary for the two groups. This method is leveraged on a smaller set of assumptions compared to linear regression (Gelman & Imbens, 2019). Locally estimated scatter plot smoothing (**loess**) is a technique for fitting an empirical curve to a scatter plot. The loess method fits a set of polynomial regression models (allowing curvature) to the data at many locations spanning the range of the x-variable (F, in this application), each based on weighting the data by proximity (Cleveland & Devlin, 1988). This set of regressions is constrained such that they produce a smooth curve across F. Loess is useful because it does not require the researcher to specify the form of the relationship between variables, and also because the uncertainty of the solution can be calculated, enabling the construction of confidence intervals and statistical tests.

One of the most important details of loess is the choice of the *bandwidth* or *span* parameter. This provides an operational definition of 'local'. A larger bandwidth allows more distant data to influence the curve, resulting in a smoother curve which ignores small-scale features in the data. A smaller bandwidth does the opposite (Jacoby, 2000).

Figure 7.6 illustrates the result of applying a loess smoother, with an appropriate choice of bandwidth, to the non-linear data (from Figure 7.4) and the interaction data (from Figure 7.5). In both cases, the true functional form is well-approximated by the loess smoother.

The estimated values at the treatment boundary (z) can be obtained in R by first applying the smoother to the treatment and control subsets of the data with the `loess()` function and then running `predict()` on the objects it creates to obtain the point estimates of the fitted curves near the boundary on the forcing variable. The `predict()` function returns an estimate of this point estimate (`$fit`) and its standard error (`$se.fit`) as components of the output. The treatment effect is calculated as the difference in predicted values near the boundary. (The predictions are made near rather than at the boundary because the `predict()` function will only return values for locations interior to the data.)

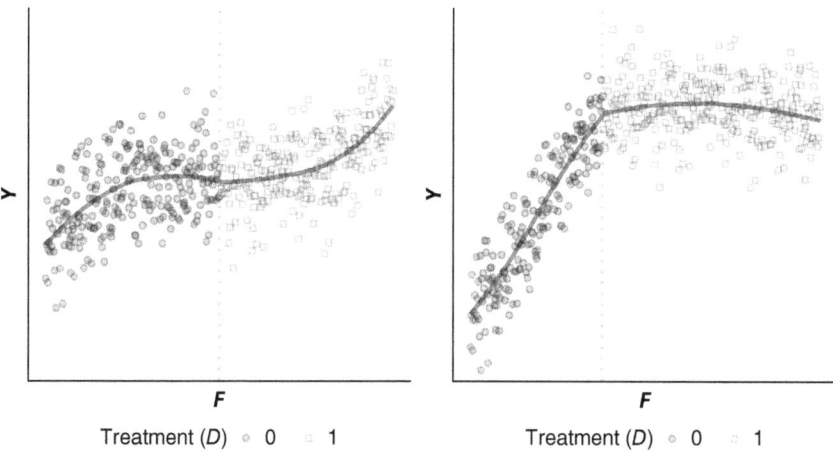

F

Treatment (D) ○ 0 □ 1

F

Treatment (D) ○ 0 □ 1

Figure 7.6 Semi-parametric locally weighted scatter plot smoothing for modelling the relationship between the response variable and the forcing variable. Left panel: Non-linear form example. Right panel: Interaction example. In both applications, the loess method provides a good representation of the data and leads to unbiased treatment effect estimates

The predict() function's output for the treated cases is as follows. The $fit value is the model-derived expected value of Y at the boundary for that group, $E (Y | D = 1, F = z)$, and $se.fit its standard error.

```
## $fit
## [1] 4.80653
##
## $se.fit
## [1] 0.3686669
##
## $residual.scale
## [1] 1.982014
##
## $df
## [1] 308.3427
```

The predict() function's output for the control cases is as follows. This is the expected value at the boundary for the control group, $E (Y | D = 0, F = z)$

```
## $fit
## [1] 4.546244
##
## $se.fit
## [1] 0.5078327
```

```
##
## $residual.scale
## [1] 2.180623
##
## $df
## [1] 184.3374
```

Assuming that the estimates are independent, the standard error of the treatment effect estimate, $E\,(Y\,|\,D = 1, F = z) - E\,(Y\,|\,D = 0, F = z)$, can be obtained by taking the square root of the sum of the squared standard errors. This is because independent errors combine in quadrature (Baird, 1962). If A and B are random variables with uncertainties ΔA and ΔB, then the uncertainty of $A - B$ is approximately $\sqrt{\Delta A^2 + \Delta B^2}$.

For the interaction example (Figure 7.5), the estimated treatment effect using locally weighted smoothing is 0.260 (0.628), 95% CI [–0.970, 1.490], approximate $p = .678$. The true treatment effect of zero is captured in the confidence interval.

The non-parameteric approach protects against the consequences of violating assumptions regarding the functional form, but this protection comes at a price. That price is the variance of the estimate, in other words, its precision. The standard errors for the treatment effect estimated via the non-parametric loess smoother will be larger than those estimated with parametric regression. The standard error of the estimated treatment effect was nearly twice as large for the loess model versus the linear regression model (0.628 vs 0.381).

The amount of efficiency loss is controlled by the bandwidth parameter. Small values of the bandwidth mean that the model-predicted values of Y at the boundary for each group are based only on the highly proximal data points. By implication, precision suffers because the inference is made on the basis of a small fraction of the data. When the bandwidth is large, distant cases make some contribution to the extrapolated value. This uses more of the data and increases precision but imposes a higher risk of misspecification bias, as real non-linear features of the data could be smoothed out of existence.

There is a bias–variance trade-off involved in the bandwidth choice, but it is not a simple trade-off such that small bandwidths always mean less bias but more variance. Very small bandwidth choices produce overfitting, in which the smoothed curve is over-responsive to meaningless noise in the data. Figure 7.7 shows the result of applying loess with an inappropriately small value of the bandwidth parameter. Local fluctuations in the data cause the smoother to suggest a discontinuity where none exists.

The overfitting is apparent in the jagged appearance of the curve. The standard error of the estimated treatment effect (1.564) using this excessively small bandwidth was over twice as large as the one estimated by the appropriate loess model (0.628).

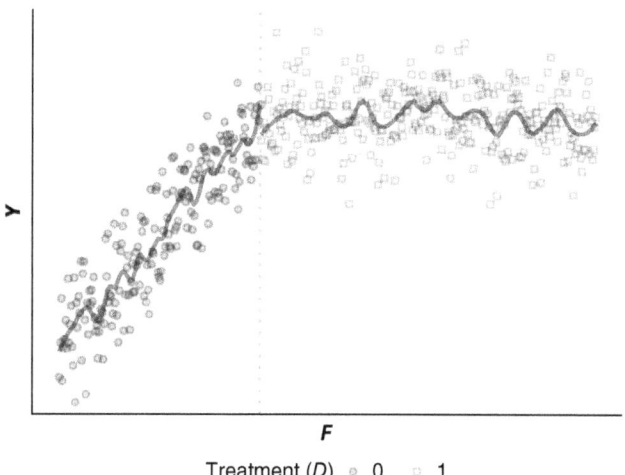

Treatment (D) ∘ 0   □ 1

Figure 7.7 Locally estimated scatter plot smoothing estimation with a small bandwidth. Small-scale fluctuations create the illusion of a discontinuity where none exists

Regardless of whether analysts choose the parametric or non-parametric method for extrapolating to the boundary, it is important that a sensitivity analysis (Saltelli et al., 2004) be presented showing how different choices of model specification or bandwidth affect estimated treatment effects. Ideally the estimate is consistent across different methods for calculating it.

Example of sharp RDD analysis

I simulated a data set of $n = 500$ for this example analysis. The data are available for download from this book's companion website. In this data set, the relationship between F and Y is non-linear, as shown in Figure 7.4. The true treatment effect is 10.0 and the cut point (z) is at $F = 1$.

Analysis with parametric regression

Table 7.1 provides the results of fitting four different regression models to the data with escalating complexity, each via R's lm() function. The models range from a simple linear model with no interactions to a complex cubic model with interactions on all the parameters. In each case, the forcing variable was centred at the cut point, enabling the main effect of D from the regression results to estimate the discontinuity at the boundary – in other words, the treatment effect.

Table 7.1 Estimated effects from parametric regression analysis of sharp RDD data

Model	Estimate (SE)	95% CI
$Y_i = b_0 + b_1(F) + b_2(D) + e_i$	5.644 (1.854)[a]	[2.011, 9.277]
$Y_i = b_0 + b_1(F) + b_2(F_i^2) + b_3(D) + e_i$	3.614 (2.126)[b]	[−0.552, 7.780]
$Y_i = b_0 + b_1(F) + b_2(F_i^2) + b_3(F_i^3) + b_4(D) + e_i$	11.974 (2.272)[c]	[7.522, 16.426]
$Y_i = b_0 + b_1(F) + b_2(F_i^2) + b_3(F_i^3) + b_4(F_i \times D) + b_5(F_i^2 \times D)$ $+ b_6(F_i^3 \times D) + b_7(D) + e_i$	11.671 (3.784)[d]	[4.254, 19.088]

Note. The true causal effect is 10.0. *F* was centred at the boundary value *z* prior to analysis. RDD = regression discontinuity design; CI = confidence interval.

[a]Model assumes linearity and no interaction. Reported parameter is b_2.

[b]Model assumes quadratic non-linearity and no interaction. Reported parameter is b_3.

[c]Model assumes cubic non-linearity and no interaction. Reported parameter is b_4.

[d]Model assumes cubic non-linearity and relaxes the assumption of no interaction. Reported parameter is b_7.

Presenting results from models that make different assumptions is a form of sensitivity analysis. It is clear in Table 7.1 that the treatment effect estimates vary widely over the models. Only the third and fourth models have captured the true treatment effect in their confidence intervals. I generated the data from a cubic model with no interaction, so the third model is the correct one in this example. Adding the interaction terms in the fourth model did not substantially change the point estimate of the treatment effect but did reduce the efficiency of the estimate considerably. The simplest model that can represent the data well is preferred. Because RDD models are fundamentally two dimensional (*F* vs *Y*), graphical techniques can be used to judge the adequacy of each one. Figure 7.8 displays plots of the four models presented in Table 7.1. These plots strongly suggest that models 3 and 4 provide reasonable summaries of the data while the first two do not.

Non-parametric estimation with the rdd package

Analysing the data with the rdd package is straightforward. The RDestimate() function uses the loess method to estimate the treatment effect, its standard error and the *p*-value. The default bandwidth is automatically determined using the Imbens–Kalyanaraman method (Imbens & Kalyanaraman, 2012), which minimises the expected mean squared error of the estimated treatment effect. This works well in most circumstances, but this can be overridden by the user if desired. A sensitivity analysis (Saltelli et al., 2004) is automatically performed by re-estimating the treatment effect with half and twice the chosen bandwidth. The sensitivity analysis shows

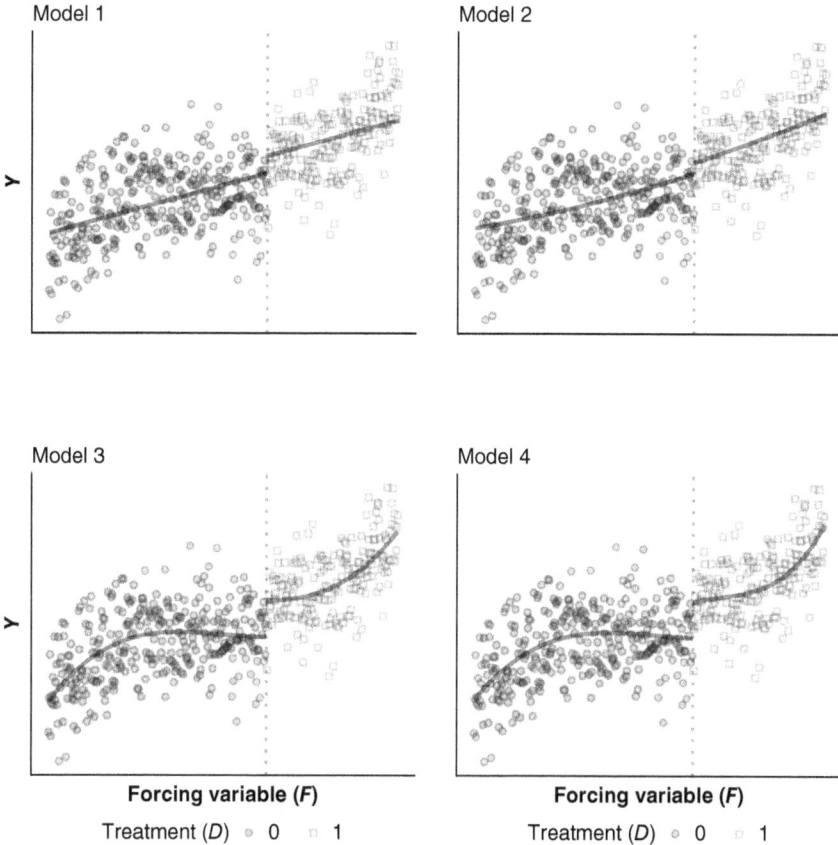

Figure 7.8 Plot of the regression solutions against the data for the models presented in Table 7.1

how the result varies over choices of bandwidth. Ideally, the three results will be quite similar, indicating that the bandwidth selection does not have much impact.

For sharp RDD, the analyst writes a formula in which the outcome is regressed on the forcing variable. The only difference in syntax for fuzzy RDD is that the treatment variable must be included in the model formula. The cut-off must be specified; otherwise the program defaults to using zero as the cut-off.

The output from running `summary()` on the object containing `RDestimate()`'s results is given below.

```
##
## Call:
## RDestimate(formula = Y ~ Forcing, data = data, cutpoint = 1)
##
## Type:
## sharp
```

```
##
## Estimates:
##              Bandwidth  Observations  Estimate  Std. Error  z value  Pr(>|z|)
## LATE         1.652      206           11.907    2.823       4.218    2.470e-05
## Half-BW      0.826      103            9.857    3.930       2.508    1.213e-02
## Double-BW    3.304      394            9.407    2.031       4.632    3.618e-06
##
## LATE         ***
## Half-BW      *
## Double-BW    ***
## ---
## Signif. codes:  0 '***' 0.001 '**' 0.01 '*' 0.05 '.' 0.1 ' ' 1
##
## F-statistics:
##              F       Num. DoF  Denom. DoF  p
## LATE         30.15   3         202         8.882e-16
## Half-BW      16.88   3          99         1.251e-08
## Double-BW    84.77   3         390         0.000e+00
```

All three of the estimated effects are similar to each other and are close to the true value of 10.0. Using the LATE (default bandwidth), the estimated treatment effect is 11.907 (2.823), 95% CI [6.374, 17.440], $p < .001$.

The effect of the various bandwidth choices on the proportion of the sample that is used, and therefore the precision, is clearly indicated by the Observations component of the RDestimate() output. The default bandwidth effectively used less than half of the data in this case.

The plot() function can be run on the RDestimate object to produce a visual depiction of the data, smoothed curves and the discontinuity, as shown in Figure 7.9. This plot is produced with base-R graphics, but publication-ready plots can be produced with the ggplot2 package.

Fuzzy RDD

In fuzzy RDD, thresholding the forcing variable is not the sole determination of treatment status (Bloom, 2012). There are exceptions to the rule – cases on either side of the boundary that do not belong there based on their values of F. Figure 7.10 displays a plot of the forcing variable versus the probability of receiving treatment. In sharp RDD, treatment is a discontinuous function of F, while in fuzzy RDD, it is a smooth function of F.

The fuzzy RDD scenario can be represented with a DAG, as shown in Figure 7.11. This DAG differs from the sharp RDD (see Figure 7.1) situation in two ways. First,

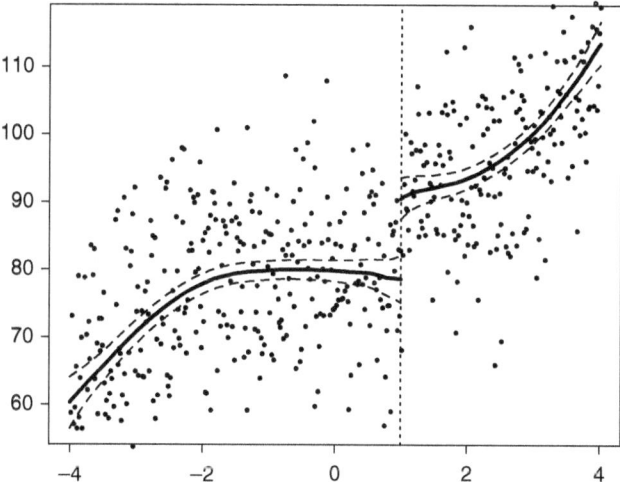

Figure 7.9 Plot of the RDestimate() function's non-parametric solution for the sharp RDD analysis

Note. RDD = regression discontinuity design.

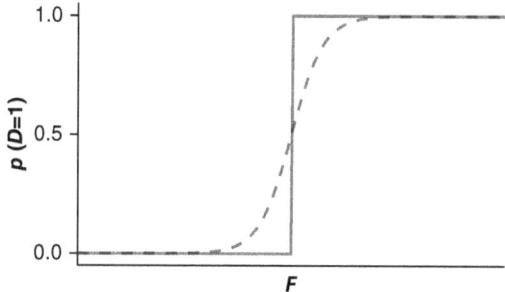

Figure 7.10 Sharp versus fuzzy RDD treatment allocation. Treatment assignment in fuzzy RDD is a probabilistic function of the forcing variable. Solid line: Sharp RDD. Dashed line: Fuzzy RDD

Note. RDD = regression discontinuity design.

an unobservable confounder U2 of the $D \rightarrow Y$ path has been added. Second, a new binary mediator, E, of the $F \rightarrow Y$ path has been added to the graph. The E variable is envisioned as *treatment encouragement*, which takes values of 0 (not encouraged) and 1 (encouraged), and it is solely determined by F. This assumption that E is completely determined by thresholding F is often quite reasonable in practice. Subjects with $F > z$ are offered treatment ($E = 1$), but some may refuse it. The probability of refusal could be related to Y by some confounder ($U2$). Similarly, subjects with $F > z$ are not offered the treatment ($E = 0$), but may obtain it through alternate means. Treatment *status* is not perfectly controlled by F, but treatment *encouragement* is.

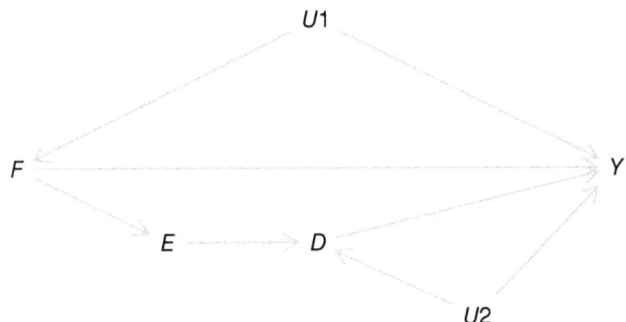

Figure 7.11 Example of a DAG for a fuzzy RDD. *F* = forcing variable; *E* = encouragement variable; *D* = treatment status. *Y* = response; *U1*, *U2* = unobserved confounders

Note. DAG= directed acyclic graph; RDD= regression discontinuity design.

Conditioning on *F* is no longer sufficient for estimating the causal effect of *D* on *Y*; doing so blocks the confounding from *U1* as well and the direct effect of *F*, but does not block the back-door path from *D* to *Y* via *U2*. In fact, no conditioning method can identify the causal effect in this situation because *U2* is unobservable. (If it were observable, in principle we could condition on it.)

Figure 7.12 displays the DAG after conditioning on *F*. All the paths emanating from *F* have been removed. This DAG reveals the strategy for identifying the treatment effect. The encouragement variable *E* has become an instrument! Instrumental variables analysis (which was described in Chapter 6) can be used to estimate the causal effect of *D* on *Y* despite the presence of *U2* (Angrist & Imbens, 1995).

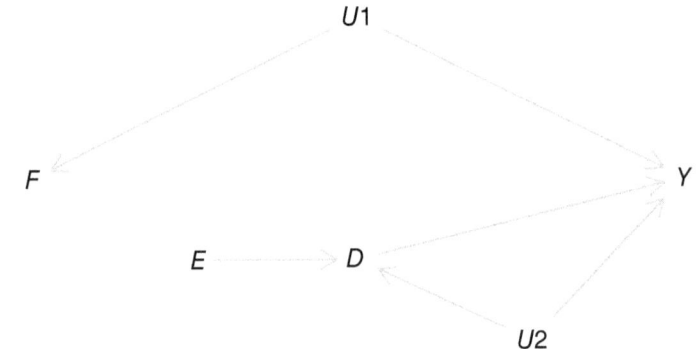

Figure 7.12 The result of conditioning on *F* in the fuzzy RDD. The encouragement variable (*E*) becomes an instrument. *F* = forcing variable; *D* = treatment status; *Y* = response; *U1*, *U2* = unobserved confounders

Note. RDD = regression discontinuity design.

Example of fuzzy RDD analysis

I simulated a data set with $n = 800$ from the DAG shown in Figure 7.11. In this data set, the relationship between F and Y is linear, and the true treatment effect is 5.0. Figure 7.13 shows a scatter plot of the data, clearly showing why this is a fuzzy RDD situation; there are exceptions to the assignment rule.

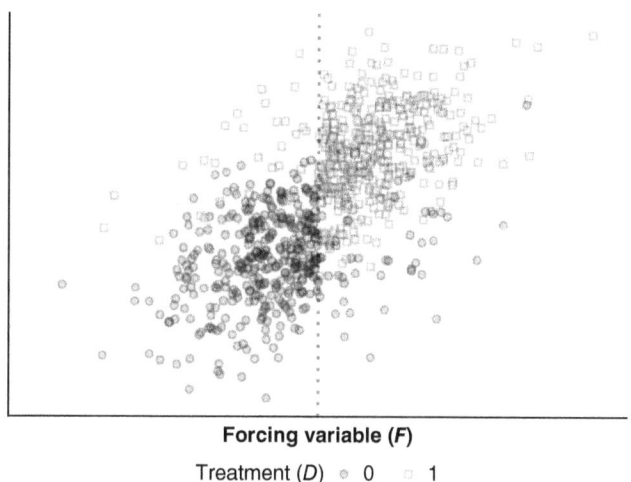

Forcing variable (*F*)

Treatment (*D*) ◦ 0   ◻ 1

Figure 7.13   Scatter plot for the fuzzy RDD data

Note. RDD = regression discontinuity design.

A cross tabulation of treatment status (*D*) by treatment encouragement (*E*) can be produced using the `table()` function, whose output is shown below. The values on the diagonal of this table are the number of cases whose treatment status corresponds with the encouragement they received based on thresholding the forcing variable. However, 79 of the 800 cases do not conform. The reason for these exceptions, according to the DAG, is the presence of the confounder *U2*.

```
##      E
## D      0    1
##    0 338   47
##    1  32  383
```

Treatment effects for the fuzzy RDD model can be estimated using two approaches. Both use 2SLS to estimate the treatment effect from the predicted treatment status based on *E*, as opposed to the actual treatment status. A parametric regression model can be fit using the `ivmodel()` function from the package AER, as was used in Chapter 6. Alternatively, a non-parametric approach based on loess can be implemented using the RDestimate() function from the rdd package. The advantages and disadvantages of these two approaches were discussed by Gelman and Imbens (2019) and Pei et al. (2014). These papers argue that parametric approach can be more efficient but also has a higher potential of being mis-specified and producing a biased result. Non-parametric smoothing approaches are preferred to parametric higher order polynomial models.

Parametric estimation using ivreg()

The ivreg() function uses a syntax in which the model formula is broken into two parts separated by a ('pipe') symbol. The regressor variables are listed on the left side of the pipe, and the instruments on the right. But exogenous predictors (those that need no IV adjustment) are listed on both sides of the pipe. Thus, the model formula supplied to ivreg() is Y~D+F|E+F. This means that the forcing variable F is an exogenous predictor that needs no IV adjustment, while the treatment variable D should be replaced with its value predicted from the encouragement variable, E. The output from this function is given below.

```
##
## Call:
## ivreg(formula = Y ~ D + F | E + F, data = data)
##
## Residuals:
##       Min        1Q    Median        3Q       Max
## -12.31742  -2.49324   0.03976   2.56343  11.03923
##
## Coefficients:
##               Estimate Std. Error t value Pr(>|t|)
## (Intercept)     0.2012     0.2943   0.684    0.494
## D               5.3422     0.5228  10.218   <2e-16 ***
## F               2.1790     0.1992  10.940   <2e-16 ***
## ---
## Signif. codes:  0 '***' 0.001 '**' 0.01 '*' 0.05 '.' 0.1 ' ' 1
##
## Residual standard error: 3.772 on 797 degrees of freedom
## Multiple R-Squared: 0.6234, Adjusted R-squared: 0.6225
## Wald test: 491 on 2 and 797 DF, p-value: < 2.2e-16
```

The simultaneous 2SLS estimate of the causal effect is 5.342, which is quite similar to the true value of 5.0. Moreover, the true value is contained in the 95% CI [4.317, 6.367].

This model assumes a linear relationship between F and Y with no interaction, but these assumptions could be relaxed by entering higher order polynomial terms of F on both sides of the pipe, as well as interactions between these components and D.

Non-parametric estimation using `RDestimate()`

Altering the `RDestimate()` syntax for fuzzy RDD is simple; the analyst simply adds the treatment status variable (*D*) to the model formula. The function automatically employs 2SLS to estimate the causal effect of the treatment. The `RDestimate()` output is as follows:

```
##
## Call:
## RDestimate(formula = Y ~ F + D, data = data, cutpoint = 0)
##
## Type:
## fuzzy
##
## Estimates:
##                 Bandwidth  Observations  Estimate  Std. Error  z value  Pr(>|z|)
## LATE            0.6868     377           5.500     0.8522      6.453    1.095e-10
## Half-BW         0.3434     204           5.286     1.2141      4.354    1.337e-05
## Double-BW       1.3737     642           5.326     0.6703      7.947    1.915e-15
##
## LATE            ***
## Half-BW         ***
## Double-BW       ***
## ---
## Signif. codes:  0 '***' 0.001 '**' 0.01 '*' 0.05 '.' 0.1 ' ' 1
##
## F-statistics:
##              F       Num. DoF  Denom. DoF  p
## LATE         70.85   3         373         2.783e-36
## Half-BW      41.20   3         200         8.913e-21
## Double-BW    157.82  3         638         1.649e-76
```

Figure 7.14 visually displays the non-parametric loess smoother and the treatment effect at the boundary.

The estimated treatment effect using the default bandwidth choice is 5.500 (0.852), 95% CI [3.829, 7.170], $p < .001$. The correct value of the treatment effect has been captured in the confidence interval.

Local average treatment effects

From the perspective of Rubin's causal model (Chapter 4), every subject has a potential outcome under treatment (Y_1) and a potential outcome under control (Y_1), only one of which can be observed. The observed values of Y when $D = 0$ are the potential

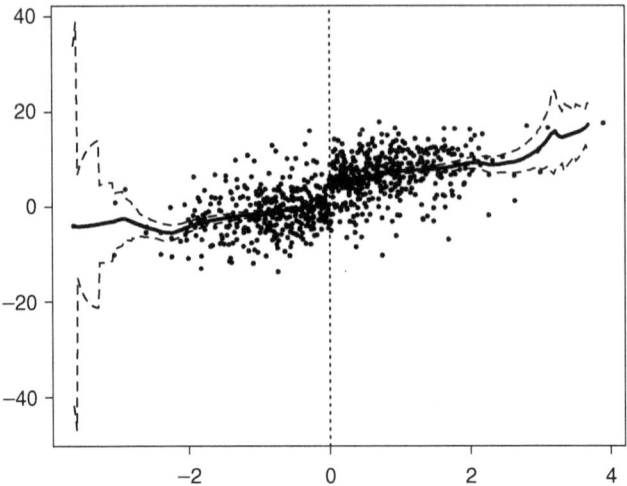

Figure 7.14 Plot of the RDestimate() function's non-parametric solution for the fuzzy RDD analysis

Note. RDD = regression discontinuity design.

outcomes $Y_0 | D = 0$, and the observed values for $D = 1$ are $Y_1 | D = 1$. The counterfactual outcomes are the unobserved values $Y_1 | D = 0$ and $Y_0 | D = 1$. Figure 7.15 displays the observed outcomes as solid lines and the counterfactuals as dotted lines. The cut point on F (at $z = 2$ in the figure) controls the location of the boundary between observing Y_0 and Y_1; the solid vertical line highlights the discontinuity (e.g. treatment effect) at this location. The boundary (z) is the only location where both of these potential outcomes can be observed, albeit from opposite sides (see Equation 7.1 and Imbens & Lemieux, 2008). The dotted vertical lines display the magnitude of the treatment effect at some different possible values of z.

The regression discontinuity at z is a LATE, which is defined as the difference in potential outcomes at z : $(Y_1 | F = z) - (Y_0 | F = z)$. Unless the functional form relating Y to F for the treatment and control groups differs only by a constant (e.g. the 'no interaction' assumption), the magnitude of the treatment effect will vary across cut point locations on the forcing variable (Hahn et al., 2001). In Figure 7.15, moving z to the right yields larger estimates of the treatment effect, while moving it to the left reduces the magnitude of the treatment effect or even causes it to become negative. The estimated treatment effect is explicitly *local*; it applies only to subjects with $F = z$. In fuzzy RDD, the LATE is conditional on both $F = z$ and compliance with treatment encouragement.

Imagine that Figure 7.15 is describing potential outcomes for the effect of gifted education programmes on academic achievement and that the forcing variable is cognitive ability. Let the potential outcome under treatment represent achievement

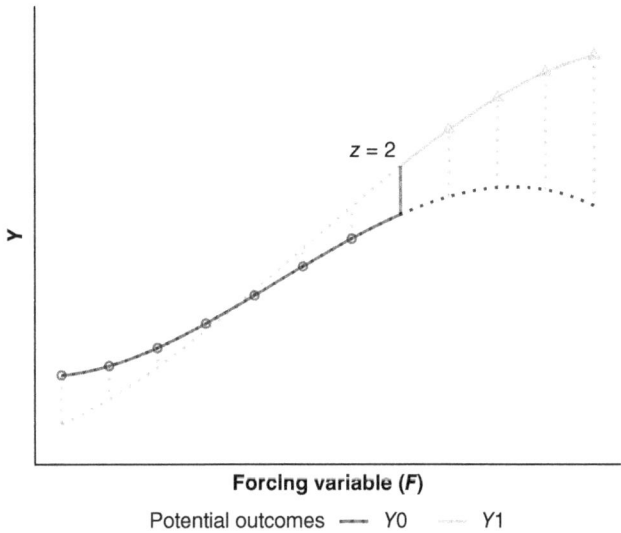

Figure 7.15 RDD estimates a LATE due to probable treatment effect heterogeneity. Solid lines: Observed outcomes using the $z = 2$ cut point. Dotted lines: Counterfactual outcomes

Note. RDD = regression discontinuity design; LATE = local average treatment effect.

in a gifted classroom and the potential outcome under control be achievement in the regular classroom. According to the figure, the relationship between achievement and cognitive ability flattens out and begin to fall in the regular classroom for $F > z$, while the relationship remains positive in the gifted classroom. Thus, a higher cut point (z) produces a larger positive treatment effect. As cognitive ability falls ($F > z$), students become less able to profit from the gifted classroom. At some point, the classrooms are equally effective, and below that the typical classroom is superior.

This property of RDD does limit the generalisability of its results. The cut point should be carefully selected such that the estimated treatment effect carries maximum relevance. But the notion of a local treatment effect can occasionally be more relevant than an average treatment effect would be for understanding the effect of targeted interventions or policies. When the appropriateness of a particular value of a cut point is in question, a local treatment effect can be the most informative treatment effect to consider.

When researchers are willing to fully commit to regression assumptions, RDD models can be used to calculate the expected LATE for any choice of z. Such applications are extraordinarily leveraged on assumptions because 'virtually' moving the cut-off by using the regression results requires making a model-based extrapolation not just to the edge of each group's observed data but beyond it for one group. The reasonableness of this extrapolation cannot be assessed even informally because doing so would require the very observations that are missing. The best that can be said is that attempting to estimate LATEs for cut-offs other than the one actually employed is highly speculative. Researchers employing non-parametric techniques to smooth the data cannot calculate expected treatment effects for alternative choices of z. Perhaps this should be regarded as a feature rather than as a shortcoming of smoothing techniques. Estimating multiple LATEs in a serious manner requires multiple studies that employ different values of z for allocating treatment status.

Assumptions of the RDD

The assumptions of the regression discontinuity design (RDD) are as follows:

- The assignment mechanism must be known.
- The potential outcomes Y_0, Y_1 must be smooth near the boundary.
- The forcing variable must not be an outcome of treatment status; it must *cause* the treatment. In other words, the DAG must be $F \to D$ rather than $D \to F$. In practice, this usually means that the forcing variable must be measured *prior* to treatment, as this rules out the possibility of reverse causation.
- The cut point is determined without reference to specific individual's scores on the forcing variable. For example, if RDD is being used to evaluate the causal effect of an educational intervention, the school cannot tailor the cut point to ensure that certain problem students 'make the cut' (so to speak).
- Individuals must not control their value of the forcing variable in order to fall on the beneficial side of the threshold. For example, individuals receiving income-based subsidies might avoid slightly higher paying jobs in order to continue receiving the subsidy.
- The only difference between individuals on either side of the cut point can be exposure to the specific treatment being evaluated and not other treatments as well. This can be a problematic assumption in educational contexts. If, for example, students with sufficiently low achievement receive remedial instruction (the focal treatment) but are also likely to be paired with a community mentor, then any discontinuity observed at the cut-off is not a valid estimate of the isolated effect of remedial instruction.

Further Reading

Gelman, A., & Imbens, G. (2017). Why high-order polynomials should not be used in regression discontinuity designs. *Journal of Business and Economic Statistics*. https://doi.org/10.1080/07350015.2017.1366909
This article argues that non-parametric smoothing is preferable to high-order polynomial models in RDD.

Imbens, G. W., & Lemieux, T. (2008). Regression discontinuity designs: A guide to practice. *Journal of Econometrics, 142*(2), 615–635. https://doi.org/10.1016/j.jeconom.2007.05.001
This paper introduces RDD and provides practical advice on how to properly use it to estimate causal effects.

Lee, D. S., & Lemieux, T. (2010). Regression discontinuity designs in economics. *Journal of Economic Literature, 48*(2), 281–355. https://doi.org/10.1257/jel.48.2.281
This paper is a general introduction to RDD for economists. It discusses conditions under which RDD estimates are likely valid or invalid and references several examples of RDD.

van der Klauuw, W. (2008). Regression-discontinuity analysis: A survey of recent developments in economics. *Labour, 22*(2), 219–245. https://doi.org/10.1111/j.1467-9914.2008.00419.x
In addition to a clear introduction to RDD, this article discusses several specification tests that may be used to detect unreliable results and also references extensions to the RDD model for situations in which the cutoff value is unknown, has multiple discontinuities, or is categorical.

8

CONCLUSION

Chapter Overview

This book has described five different methods for assessing causation: (1) randomisation of the focal variable, (2) conditioning on confounders using covariate adjustment, (3) non-parametric conditioning on confounders (e.g. via stratification, matching or weighting) aided by propensity scores, (4) IVA and (5) RDDs. Randomisation was only discussed implicitly since this book generally considers the more difficult cases of observational or quasi-experimental designs.

The most potent knowledge a researcher can possess is an understanding of how treatment assignment occurred (or more generally, how the focal variable was allocated). The better this is understood, the more reliably causal inferences can be made. Maximal certainty is obtained when the researcher assigned the focal variable to the subjects, compliance is perfect and the subjects could not obtain the treatment via any other avenue. This ensures that those subjects who were assigned to treatment actually experience the treatment, and those assigned to control do not. The least certainty occurs when the researcher has no control over assignment, and there is no body of well-tested theory or empirical findings to inform speculation about how the X variable was apportioned.

The techniques in this book can therefore be placed on a continuum corresponding to knowledge of treatment assignment. If researchers can control treatment assignment, they can assign it based on randomisation or on the basis of a well-measured quantitative variable. This first condition is an experiment, the latter one leads to the regression discontinuity design. Deciding between these approaches is mostly a question of external validity and precision rather than the reliability of causal inference, so long as the relationship between the forcing variable and the outcome is relatively simple. Both approaches work well, but the researcher must consider to whom they wish to make inferences. The average treatment effect estimated by randomisation describes the averaged effect over the general population, while the local average treatment effect estimated by RDD is specific to those subjects whose value of the forcing variable is near the boundary. As indicated in Chapter 7, there are certain classes of targeted interventions (e.g. gifted education programmes and special education classes) that are only intended to benefit people with certain characteristics. In such cases, knowing the average treatment effect (or the average treatment effect for the treated) is informative for understanding the programme's effect, but the impact of the intervention on students near the cut point is also crucial for understanding whether cut point is reasonable or defensible. Experiments offer greater precision and statistical power for a fixed sample size since the entire sample is equally informative, whereas in the RDD, observations lying far from the threshold have little value.

In the moderate range of knowledge of the treatment assignment process, researchers have a good chance of being able to articulate a reasonably complete and correct causal DAG. Once obtained and verified against the data (via the DAGitty package's localTests function, the DAG can be analysed to identify those variables that must be conditioned on (a sufficient adjustment set) in order to implement a conditioning-based strategy for causal inference. The set of necessary variables can be surprisingly small even when highly complex DAGs are analysed, in my experience often consisting of only three to five variables. Once such an adjustment set is identified, researchers must decide whether to use a parametric or a non-parametric conditioning strategy. These decisions are technical in nature and have, frankly, much lower stakes than the determination of which variables to include in the adjustment set. All of the conditioning methods can produce unbiased and consistent estimates when their assumptions are met. Covariate adjustment offers the best precision, but it is the most dependent on assumptions of functional form and is the most difficult to appraise for quality. Still, it works reasonably well much of the time. Large sample sizes make it easier to justify adopting a non-parametric strategy, trading off some precision in favour of the potential for greater accuracy and the availability of useful diagnostics related to common support and covariate balancing.

When researchers have little understanding of the assignment process, the prospect of making correct conditioning decisions is bleak, so that any conditioning-based method is likely to fail. The same situation arises when knowledge is plentiful but valid and reliable measures of the important variables are not. In this case, so long as researchers can identify at least one factor that causes the exposure variable to change, but does not directly impact the outcome, causal inference is possible using instrumental variables analysis. For example, one does not need to understand (and assess) the myriad factors that cause someone to become a smoker; they need only understand that the price of cigarettes is one of them and that the price of cigarettes cannot affect health except by altering smoking behaviour. This is a much more obtainable goal in many circumstances. While instrumental variables methods require much larger sample sizes than equivalent experiments to achieve a useful level of precision, the widespread and increasing availability of huge administrative data sets makes this somewhat less of a concern.

Some practical advice

Disclose your DAG

In the field in which I was trained (eductional psychology), a large majority of research projects involves fitting regression models (or their analogues) to data in order to estimate the effect of some X on some Y after controlling for covariates.

But almost none of these papers provide a DAG justifying the selection of these covariates, mostly because DAGs were not part of the training for the scholars in my field. Decisions about which covariates to include seem to be made intuitively, following 'rules of thumb' that are passed down from advisor to student (e.g. 'always control for race, sex, and gender!'), or worst of all, by deciding which covariates to include based on the bivariate correlations that they have with the outcome (Simmering, 2014). None of these strategies are likely to select an adjustment set that satisfies the back-door criterion (or equivalently, **ignorable treatment assignment**) that is necessary for the valid estimation of causal relationships.

It is important that researchers take the time to construct and analyse a DAG to inform these critical decisions. It is equally important to disclose their working DAG in research reports (ideally as Figure 1, but at least in supplemental materials) not only to provide justification for the choices they made but also to make their assumptions transparent and expose them to criticism. This will greatly facilitate the accumulation of scientific knowledge.

Test your DAG

A DAG should not just be assumed to be true; it should be evaluated. Researchers should test the conditional indepedencies in their data against those that are implied by the DAG (the localTests() function in the DAGitty package automates this) to reject DAGs that cannot be correct. These tests produce asymmetrical evidence, in that a failed result is strong evidence of an incorrect DAG, but a passing result is very weak evidence of a correct DAG. Still, it is surprising how many proposed DAGs end up failing these tests. They are quite useful, if imperfect.

Once a passing DAG has been identified, researchers should consider the plausibility of equivalent DAGs, which can be generated using the equivalentDAGs() function in DAGitty. When these cannot be rejected on logical or theoretical grounds, a sensitivity analysis (Saltelli et al., 2004) should be considered to examine how the estimated causal effect varies over different adjustment sets. Ideally, the estimates across multiple models will be consistent.

Sensitivity analysis

A sensitivity analysis should be considered whenever a researcher has to make arbitrary or relatively uninformed decisions that can affect the result. Ideally, these decisions should have little impact on the resulting causal estimate; in this case, the decision is not a critical one. Examples of decisions whose impact could be examined through sensitivity analysis include the following:

- Regression versus propensity score analysis
- Subclassification versus matching versus weighting
- Truncated versus untruncated IPTW weights
- Adjustment sets derived from equivalent, data-consistent DAGs
- The number of strata
- Matching algorithms
- The smoothing bandwidth in non-parametric RDD

Sensitivity analysis simply involves repeating the analysis over different decisions to determine the degree to which the result is sensitive to these decisions; a robust result is one that does not vary across arbitrary analytic choices. Relatively new methods known as *multiverse analysis* (Steegen et al., 2016) and *specification curves* (Simonsohn et al., 2019) extend the concept of sensitivity analysis to a much wider set of decisions.

Don't ignore precision

It is tempting to think of bias (or consistency) as the only factor that matters, but the 'goodness' or utility of an estimator is actually a function of its MSE – how 'wrong' it is likely to be, on average – which is affected by its bias *and* its precision. Precision matters a great deal, and in many cases, unbiased estimators do not have the smallest MSE. Sometimes a biased but low-variance estimator is better (James et al., 2013).

Don't fool yourself

Richard Feynman (1974) famously quipped, 'The first principle is that you must not fool yourself – and you are the easiest person to fool'. It is unfortunately quite easy to fool yourself and others with the results of statistical analysis conducted with excessive freedom to iterate over many analytical models. It becomes quite easy to consciously or unconsciously select the version of the analysis that you adopt and report based on the favourability of the result. Such '*p*-hacking' (Wicherts et al., 2016) is thought to have played a major role in the replication crisis, which at the time of this writing is adversely affecting many disciplines of the behavioural and biomedical sciences (Shrout & Rodgers, 2018).

I recommend that researchers learn about and incorporate open science practices into their scientific workflow, especially the act of preregistration (Nosek et al., 2018). Preregistration is a time-stamped detailed research plan that is completed before a study takes place, documenting the a priori hypotheses, procedure, methods and statistical analysis plan. The preregistration document can specify a flow chart of if–then decisions to be evaluated once the data are in hand; it can document a decision-making process as much as a concrete set of decisions. The purpose of preregistration is to create a verifiable distinction between the confirmatory and exploratory aspects of a study and to prevent or reduce the undisclosed freedom that researchers can apply to their analysis. Preregistration does not prevent exploratory data analysis, it simply requires that exploratory findings be appropriately described as such.

What to learn next

The 'further reading' recommendations in each chapter provide a good place to start for readers who wish to continue learning about these topics. This book's References section contains some of the best available work on causality and related subjects, though I have doubtlessly overlooked many excellent contributions. Beyond that, I offer readers the following recommendations regarding where to go next.

Campbell and Stanley's versus Rubin's perspectives on causation

A 2010 special issue of *Psychological Methods* (volume 15, issue 1) was devoted to the discussion, comparison and integration of these two perspectives (Maxwell, 2010). It is an excellent resource that I highly recommend.

More about DAGs

The literature on DAGs is vast and fascinating, and this book has only introduced a small subset of the concept and application of graphical causal models. Interested readers will enjoy Rohrer (2018), Elwert (2013) and then Pearl et al. (2016) as an entry into the more advanced writing on DAGs, such as Pearl (2009). The tutorials, examples and linked papers on the dagitty.net website are particularly helpful. Pearl and MacKenzie's (2018) *The Book of Why* is an introduction to the Pearlian worldview on causal inference written for a lay audience and helps situate DAGs into a larger philosophy of the role of data, the value of experiments and the contribution of work on causal inference to the larger fields of statistics, artificial intelligence and the philosophy of science.

Principal stratification

One rule that can be justified by DAGs is that one should not condition on a post-treatment variable (generally, any variable caused by X), as these are either mediators or colliders. But there are circumstances in which this conditioning is either of inter-est or unavoidable. One common example is in an experiment with imperfect com-pliance, in which the researcher is interested in estimating the effect of the treatment for those subjects that actually complied with their treatment assignment. A naive comparison of means between the treatment and control subjects who appeared to comply with their assigned treatment status does not estimate a valid causal effect because compliance status was not randomly assigned. Principal stratification, also known in this context as CACE models, is an application of instrumental variables methods for estimating treatment effects in situations like these (Frangakis & Rubin, 2002). The literature is quite approachable for readers who are comfortable with Rubin's causal model notation and understand instrumental variables.

Fixed effects

Fixed effects methods rely on longitudinal data to remove potentially unobservable confounding bias. In longitudinal studies, variables can be categorised as time-stable (e.g. race) or time-varying (e.g. hours of sleep per night). Fixed effects models remove the effects time-stable confounding, allowing the estimation of the causal effect of time-varying exposure variables. The basic intuition is that individuals can serve as their own controls (Allison, 2009). The simplest fixed effect model is the well-known difference-in-differences estimator (Abadie, 2005), but the approach may be extended to cross-lagged models of reciprocal causation over many measurement occasions. The identifying assumption of fixed effects models is that there is no time-varying confounding, which can sometimes be implausible, and somewhat limits their applicability.

Closing

In closing, I hope that you have enjoyed this book and found this information to be useful, inspiring and clearly communicated. Every area of social research can benefit from careful application of the methods of causal inference. While the challenge is immense, so are the rewards.

GLOSSARY

ATE: The average treatment effect; the average causal effect over the whole sample.

ATT: The average treatment effect for the treated; the average causal effect for the subjects who received the treatment.

ATU: The average treatment effect for the untreated; the average causal effect for the subjects who did not receive the treatment. This is sometimes called the *average treatment effect for control* (ATC).

Back-door criterion: In the DAG framework, a condition that allows causal effects to be estimated from observational data. It is equivalent to *ignorable treatment assignment*.

Bias: The difference between the expected value of a statistic and the population parameter it is intended to estimate. Confounding is a major cause of bias.

Boosting: A machine learning algorithm in which an ensemble of simple models is used to predict the value or category of data.

Broken experiment: An experiment with imperfect compliance with treatment, differential dropout across conditions, or some other factor that renders the treatment and control groups non-comparable in the absence of treatment.

Caliper: The maximum tolerable difference between values in order for two cases to be matched with one another.

Causal effect: The change in Y that would occur given a manipulation or alteration of X.

Chain. A configuration or sub-configuration within a DAG in which causation is transmitted from one variable to the next in a consistent direction without splitting. For example, $A \rightarrow B \rightarrow C$ is a chain.

Classification tree: A non-parametric method for predicting class membership from data based on recursive partitioning.

Common support: The range of propensity scores in which there is a sufficient density of both treatment and control cases to justify estimating causal effects.

Conditioning: The act of incorporating additional information into an analysis.

Conditional independence: When a pair of variables are rendered statistically independent due to an act of conditioning.

Confounder: A variable which causes both the exposure/treatment variable and the outcome, inducing a spurious correlation between them.

Counterfactual: The outcome variable under a condition that did not occur.

DAG: A directed acyclic graph or causal diagram.

Direct effect: The causal inference of one variable on another that is transmitted directly, without passing through any intervening variables.

Endogenous variable: A variable which is caused by another variable within the framework of a DAG. An endogenous variable will have at least one arrow directed toward it.

Endogeneity: A correlation between a model's errors and one or more of its predictor variables, resulting in biased and inconsistent estimates. Endogeneity results from confounding.

Exogenous variable: A variable which is not caused by any other variable within the framework of a DAG, or whose causes are deemed irrelevant to the question at hand.

Experiment: A research design in which subjects are randomly assigned to treatment conditions.

External validity: The extent to which results are generalised beyond the research setting to a population of interest.

Forcing variable: The quantitative variable that dictates (or at least strongly influences) treatment assignment in a regression discontinuity design.

Fork: A structure or substructure in a DAG in which causation is transmitted along multiple pathways emanating from a common ancestor. For example, $A \leftarrow B \rightarrow C$ is a fork originating with B.

Ignorable treatment assignment: In Rubin's causal model, the condition in which the potential outcomes are independent of treatment assignment, allowing treatment effects to be estimated from observational data.

Instrument (or instrumental variable): A variable that causes the exposure and has no causal effect on the outcome except via the exposure.

Internal validity: The degree to which a study supports causal inference.

Kernel density estimation: A non-parametric method for estimating the probability distribution of a variable.

LATE: A local average treatment effect, which is specific to a particular subgroup.

Loess: Locally estimated scatter plot smoothing; a non-parametric technique for estimating and visualising the relationship between variables.

Measurement error: Noise in observed scores which causes them to deviate from the true scores. In the classical test theory model, measurement errors are assumed to be independent of the true scores.

Observed score: The values resulting from a measurement operation, consisting of the true score plus measurement error.

Overfitting: A model that has become over-responsive to change features in the sample (or training data), causing it to perform poorly when applied to out-of-sample data.

Potential outcomes: The values of the response variable that for a given subject would have been observed under the treatment and control conditions. Only one of these is observable, the other is a *counterfactual*.

Precision: The uncertainty of a statistical estimate as typically measured by its standard error or variance. Less uncertainly = smaller standard error/variance = more precision.

Propensity score: The predicted probability of receiving treatment based on covariates.

Reliability: The proportion of variance in a measurement that is not measurement error (e.g. noise).

Threat to validity: A potential alternative non-causal explanation for a study's result.

Total effect: The total causal influence of one variable on another, including both direct and indirect pathways.

Treatment effect: See *causal effect*.

True score: The underlying true value of the quantity to be measured. According to classical test theory, these scores cannot be observed, because all measurements are contaminated with some degree of measurement error.

Unobservable: A variable that was not or cannot be measured.

Check out the next title in the collection: *Big Data Mining and Complexity*, for guidance on Data Mining and 'Big Data'.

REFERENCES

Abadie, A. (2005). Semiparametric difference-in-differences estimators. *Review of Economic Studies, 72*(1), 1–19. https://doi.org/10.1111/0034-6527.00321

Allison, P. D. (2009). *Fixed effects regression models* (Vol. *160*). Sage. https://doi.org/10.4135/9781412993869

Angrist, J., & Imbens, G. (1995). Identification and estimation of local average treatment effects. *Econometrica, 62*(2), 467–475. https://doi.org/10.2307/2951620

Angrist, J. D., Imbens, G. W., & Rubin, D. B. (1996). Identification of causal effects using instrumental variables. *Journal of the American Statistical Association, 91*(434), 444–455. https://doi.org/10.2307/2291629

Angrist, J. D., & Krueger, A. B. (2001). Instrumental variables and the search for identification: From supply and demand to natural experiments. *Journal of Economic Perspectives, 15*(4), 69–85. https://doi.org/10.1257/jep.15.4.69

Apel, R. J., & Sweeten, G. (2010). Propensity score matching in criminology and criminal justice. In A. Piquero & D. Weisburd (Eds.), *Handbook of quantitative criminology* (pp. 543–562). Springer. https://doi.org/10.1007/978-0-387-77650-7_26

Auguie, B., & Antonov, A. (2017). *gridExtra: Miscellaneous functions for "grid" graphics.* https://CRAN.R-project.org/package=gridExtra

Austin, P. C. (2011). An introduction to propensity score methods for reducing the effects of confounding in observational studies. *Multivariate Behavioral Research, 46*(3), 399–424. https://doi.org/10.1080/00273171.2011.568786

Austin, P. C., & Stuart, E. A. (2015). Moving towards best practice when using inverse probability of treatment weighting (IPTW) using the propensity score to estimate causal treatment effects in observational studies. *Statistics in Medicine, 34*(28), 3661–3679. https://doi.org/10.1002/sim.6607

Baiocchi, M., Cheng, J., & Small, D. S. (2014). Instrumental variable methods for causal inference. *Statistics in Medicine, 33*(13), 2297–2340. https://doi.org/10.1002/sim.6128

Baird, D. C. (1962). *Experimentation: An introduction to measurement theory and experiment design.* Prentice Hall.

Barnard, J., Frangakis, C. E., Hill, J. L., & Rubin, D. B. (2003). Principal stratification approach to broken randomized experiments: A case study of school choice vouchers in New York City. *Journal of the American Statistical Association, 98*(462), 299–323. https://doi.org/10.1198/016214503000071

Bascle, G. (2008). Controlling for endogeneity with instrumental variables in strategic management research. *Strategic Organization, 6*(3), 285–327. https://doi.org/10.1177/1476127008094339

Baser, O. (2006). Too much ado about propensity score models? Comparing methods of propensity score matching. *Value in Health, 9*(6), 377–385. https://doi.org/10.1111/j.1524-4733.2006.00130.x

Bauer, D. J., & Curran, P. J. (2005). Probing interactions in fixed and multilevel regression: Inferential and graphical techniques. *Multivariate Behavioral Research, 40*(3), 373–400. https://doi.org/10.1207/s15327906mbr4003_5

Bellman, R. E. (1961). *Adaptive control processes: A guided tour.* Princeton University Press. https://doi.org/10.1515/9781400874668

Bembom, O., & van der Laan, M. J. (2008). *Data-adaptive selection of the truncation level for inverse-probability-of-treatment-weighted estimators* (Working Paper No. 230). UC Berkeley Division of Biostatistics. https://biostats.bepress.com/ucbbiostat/paper230/

Berk, R. A. (2004). *Regression analysis: A constructive critique* (Vol. 11). Sage. https://doi.org/10.4135/9781483348834

Bloom, H. S. (2012). Modern regression discontinuity analysis. *Journal of Research on Educational Effectiveness, 5*(1), 43–82. https://doi.org/10.1080/19345747.2011.578707

Boef, A. G., Dekkers, O. M., Vandenbroucke, J. P., & le Cessie, S. (2014). Sample size importantly limits the usefulness of instrumental variable methods, depending on instrument strength and level of confounding. *Journal of Clinical Epidemiology, 67*(11), 1258–1264. https://doi.org/10.1016/j.jclinepi.2014.05.019

Bollen, K. A. (2012). Instrumental variables in sociology and the social sciences. *Annual Review of Sociology, 38*, 37–72. https://doi.org/10.1146/annurev-soc-081309-150141

Bollen, K. A. (2014). *Structural equations with latent variables* (Vol. 210). Wiley. https://doi.org/10.1002/9781118619179

Borenstein, M., Hedges, L. V., Higgins, J. P., & Rothstein, H. R. (2011). *Introduction to meta-analysis.* Wiley.

Breiman, L. (1996). Bagging predictors. *Machine Learning, 24*(2), 123–140. https://doi.org/10.1023/A:1018054314350

Breiman, L., Friedman, J. H., Olshen, R. A., & Stone, C. J. (1984). *Classification and regression trees.* Wadsworth.

Brookhart, M. A., Stürmer, T., Glynn, R. J., Rassen, J., & Schneeweiss, S. (2010). Confounding control in healthcare database research: Challenges and potential approaches. *Medical Care, 48*(6), S114–S120. https://doi.org/10.1097/MLR.0b013e3181dbebe3

Bullock, J. G., Green, D. P., & Ha, S. E. (2010). Yes, but what's the mechanism? (Don't expect an easy answer). *Journal of Personality and Social Psychology, 98*(4), 550–558. https://doi.org/10.1037/a0018933

Burgess, S., Small, D. S., & Thompson, S. G. (2017). A review of instrumental variable estimators for Mendelian randomization. *Statistical Methods in Medical Research, 26*(5), 2333–2355. https://doi.org/10.1177/0962280215597579

Campbell, D. T., & Stanley, J. C. (1966). *Experimental and quasi-experimental designs for research.* Houghton Mifflin.

Canty, A., & Ripley, B. (2020). *boot: Bootstrap functions* (Originally by Angelo Canty for S). https://CRAN.R-project.org/package=boot

Cleveland, W. S., & Devlin, S. J. (1988). Locally weighted regression: An approach to regression analysis by local fitting. *Journal of the American Statistical Association, 83*(403), 596–610. https://doi.org/10.1080/01621459.1988.10478639

Cochran, W. G. (1968). The effectiveness of adjustment by subclassification in removing bias in observational studies. *Biometrics, 24*(2), 295–313. https://doi.org/10.2307/2528036

Crocker, L. M., & Algina, J. (1986). *Introduction to classical and modern test theory.* Holt, Rinehart & Winston.

Crump, R. K., Hotz, V. J., Imbens, G. W., & Mitnik, O. A. (2009). Dealing with limited overlap in estimation of average treatment effects. *Biometrika, 96*(1), 187–199. https://doi.org/10.1093/biomet/asn055

Cumming, G. (2013). *Understanding the new statistics: Effect sizes, confidence intervals, and meta-analysis.* Routledge. https://doi.org/10.4324/9780203807002

Dawid, A. P. (2010). Seeing and doing: The Pearlian synthesis. In R. Dechter, H. Geffner, & J. Y. Halpern (Eds.), *Heuristics, probability and causality: A tribute to Judea Pearl* (pp. 309–325). http://bayes.cs.ucla.edu/TRIBUTE/festschrift-complete.pdf

Dayton, A., Exner, E. C., Bukowy, J. D., Stodola, T. J., Kurth, T., Skelton, M., Greene, A. S., & Cowley, A. W., Jr. (2016). Breaking the cycle: Estrous variation does not require increased sample size in the study of female rats. *Hypertension, 68*(5), 1139–1144. https://doi.org/10.1161/HYPERTENSIONAHA.116.08207

Deaton, A., & Cartwright, N. (2018). Understanding and misunderstanding randomized controlled trials. *Social Science & Medicine, 210*, 2–21. https://doi.org/10.1016/j.socscimed.2017.2.005

DeGroot, M. H., & Schervish, M. J. (2002). *Probability and statistics* (3rd ed.). Addison-Wesley.

de la Cuesta, B., & Imai, K. (2016). Misunderstandings about the regression discontinuity design in the study of close elections. *Annual Review of Political Science, 19*, 375–396. https://doi.org/10.1146/annurev-polisci-032015-010115

de Vries, A., & Ripley, B. D. (2020). *ggdendro: Create dendrograms and tree diagrams using "ggplot2"*. https://CRAN.R-project.org/package=ggdendro

Didelez, V., & Sheehan, N. (2007). Mendelian randomization as an instrumental variable approach to causal inference. *Statistical Methods in Medical Research, 16*(4), 309–330. https://doi.org/10.1177/0962280206077743

Dienes, Z. (2008). *Understanding psychology as a science: An introduction to scientific and statistical inference*. Red Globe Press.

Dimmery, D. (2016). *rdd: Regression discontinuity estimation*. https://CRAN.R-project.org/package=rdd

Ding, P., & Li, F. (2018). Causal inference: A missing data perspective. *Statistical Science, 33*(2), 214–237. https://doi.org/10.1214/18-STS645

Dowle, M., Srinivasan, A., Gorecki, J., Chirico, M., Stetsenko, P., Short, T., Lianoglou, S., Antonyan, E., Bonsch, M., Parsonage, H., Ritchie, S., Ren, K., Tan, X., Saporta, R., Seiskari, O., Dong, X., Lang, M., Iwasaki, W., Wenche, S., . . . Eddelbuettel, D. (2020). *data.table: Extension of "data.frame"*. https://CRAN.R-project.org/package=data.table

Elwert, F. (2013). Graphical causal models. In S. Morgan (Ed.), *Handbook of causal analysis for social research* (pp. 245–273). Springer. https://doi.org/10.1007/978-94-007-6094-3_13

Elwert, F., & Winship, C. (2014). Endogenous selection bias: The problem of conditioning on a collider variable. *Annual Review of Sociology, 40*, 31–53. https://doi.org/10.1146/annurev-soc-071913-043455

Evans, W. N., & Ringel, J. S. (1999). Can higher cigarette taxes improve birth outcomes? *Journal of Public Economics, 72*(1), 135–154. https://doi.org/10.1016/S0047-2727(98)00090-5

Feynman, R. P. (1974). Cargo cult science. *Engineering and Science, 37*(7), 10–13.

Fisher, R. A. (1935). The logic of inductive inference. *Journal of the Royal Statistical Society, 98*(1), 39–82. https://doi.org/10.2307/2342435

Fisher, R. A. (1936). Design of experiments. *British Medical Journal, 1*(3923), 554–554. https://doi.org/10.1136/bmj.1.3923.554-a

Fox, J. (1997). *Applied regression analysis, linear models, and related methods*. Sage.

Frangakis, C. E., & Rubin, D. B. (2002). Principal stratification in causal inference. *Biometrics, 58*(1), 21–29. https://doi.org/10.1111/j.0006-341X.2002.00021.x

Friedman, J. H. (2002). Stochastic gradient boosting. *Computational Statistics & Data Analysis*, *38*(4), 367–378. https://doi.org/10.1016/S0167-9473(01)00065-2

Funk, M. J., Westreich, D., Wiesen, C., Stürmer, T., Brookhart, M. A., & Davidian, M. (2011). Doubly robust estimation of causal effects. *American Journal of Epidemiology*, *173*(7), 761–767. https://doi.org/10.1093/aje/kwq439

Gelman, A., & Imbens, G. (2019). Why high-order polynomials should not be used in regression discontinuity designs. *Journal of Business & Economic Statistics*, *37*(3), 447–456. https://doi.org/10.1080/07350015.2017.1366909

Greenland, S. (2000). An introduction to instrumental variables for epidemiologists. *International Journal of Epidemiology*, *29*(4), 722–729. https://doi.org/10.1093/ije/29.4.722

Greenland, S., Pearl, J., & Robins, J. M. (1999). Causal diagrams for epidemiologic research. *Epidemiology*, *10*(1), 37–48. https://doi.org/10.1097/00001648-199901000-00008

Greenwell, B., Boehmke, B., Cunningham, J., & GBM Developers. (2020). *gbm: Generalized boosted regression models*. https://CRAN.R-project.org/package=gbm

Gu, X. S., & Rosenbaum, P. R. (1993). Comparison of multivariate matching methods: Structures, distances, and algorithms. *Journal of Computational and Graphical Statistics*, *2*(4), 405–420. https://doi.org/10.2307/1390693

Guo, S., & Fraser, M. W. (2015). *Propensity score analysis* (2nd ed.). Sage.

Hahn, G. J. (1977). The hazards of extrapolation in regression analysis. *Journal of Quality Technology*, *9*(4), 159–165. https://doi.org/10.1080/00224065.1977.11980791

Hahn, J., Todd, P., & Van der Klaauw, W. (2001). Identification and estimation of treatment effects with a regression-discontinuity design. *Econometrica*, *69*(1), 201–209. https://doi.org/10.1111/1468-0262.00183

Heckman, J. J. (1995). *Randomization as an instrumental variable* (NBER Working Paper No. 184). National Bureau of Economic Research. https://doi.org/10.3386/t0184

Helmreich, J. E., & Pruzek, R. M. (2012). *PSAgraphics: Propensity score analysis graphics*. https://CRAN.R-project.org/package=PSAgraphics

Hernán, M. A., Hernández-Díaz, S., & Robins, J. M. (2004). A structural approach to selection bias. *Epidemiology*, *15*(5), 615–625. https://doi.org/10.1097/01.ede.0000135174.63482.43

Hernán, M. A., & Robins, J. M. (2006). Instruments for causal inference: An epidemiologist's dream? *Epidemiology*, *17*(4), 360–372. https://doi.org/10.1097/01.ede.0000222409.00878.37

Ho, D. E., Imai, K., King, G., & Stuart, E. A. (2007). Matching as nonparametric preprocessing for reducing model dependence in parametric causal inference. *Political Analysis*, *15*(3), 199–236. https://doi.org/10.1093/pan/mpl013

Ho, D. E., Imai, K., King, G., Stuart, E., & Whitworth, A. (2018). *MatchIt: Nonparametric preprocessing for parametric causal inference.* https://CRAN.R-project.org/package=MatchIt

Hothorn, T., Bretz, F., Westfall, P., Heiberger, R. M., Schuetzenmeister, A., & Scheibe, S. (2019). *multcomp: Simultaneous inference in general parametric models.* https://CRAN.R-project.org/package=multcomp

Huber, M., Lechner, M., & Steinmayr, A. (2015). Radius matching on the propensity score with bias adjustment: Tuning parameters and finite sample behaviour. *Empirical Economics, 49*(1), 1–31. https://doi.org/10.1007/s00181-014-0847-1

Imbens, G. W., & Kalyanaraman, K. (2012). Optimal bandwidth choice for the regression discontinuity estimator. *Review of Economic Studies, 79*(3), 933–959. https://doi.org/10.1093/restud/rdr043

Imbens, G. W., & Lemieux, T. (2008). Regression discontinuity designs: A guide to practice. *Journal of Econometrics, 142*(2), 615–635. https://doi.org/10.1016/j.jeconom.2007.05.001

Jacoby, W. G. (2000). Loess: A nonparametric, graphical tool for depicting relationships between variables. *Electoral Studies, 19*(4), 577–613. https://doi.org/10.1016/S0261-3794(99)00028-1

James, G., Witten, D., Hastie, T., & Tibshirani, R. (2013). *An introduction to statistical learning* (Vol. *112*). Springer. https://doi.org/10.1007/978-1-4614-7138-7

Johnson, P. E., & Grothendieck, G. (2019). *rockchalk: Regression estimation and presentation.* https://CRAN.R-project.org/package=rockchalk

Kang, H., Jiang, Y., Zhao, Q., & Small, D. (2020). *ivmodel: Statistical inference and sensitivity analysis for instrumental variables model.* https://CRAN.R-project.org/package=ivmodel

Kim, D., Baum, C. F., Ganz, M. L., Subramanian, S., & Kawachi, I. (2011). The contextual effects of social capital on health: A cross-national instrumental variable analysis. *Social Science & Medicine, 73*(12), 1689–1697. https://doi.org/10.1016/j.socscimed.2011.09.019

Kish, L. & Anderson, D. W. (1978). Multivariate and multipurpose stratification. *Journal of the American Statistical Association, 74*(361), 24–34. https://doi.org/10.2307/2286511

Kitagawa, T. (2015). A test for instrument validity. *Econometrica, 83*(5), 2043–2063. https://doi.org/10.3982/ECTA11974

Kleiber, C., & Zeileis, A. (2020). *AER: Applied econometrics with R.* https://CRAN.R-project.org/package=AER

Lakens, D. (2013). Calculating and reporting effect sizes to facilitate cumulative science: A practical primer for *t*-tests and ANOVAs. *Frontiers in Psychology, 4*, Article 863. https://doi.org/10.3389/fpsyg.2013.00863

Law, G. R., Green, R., & Ellison, G. T. (2012). Confounding and causal path diagrams. In Y.-K. Tu & D. C. Greenwood (Eds.), *Modern methods for epidemiology* (pp. 1–13). Springer. https://doi.org/10.1007/978-94-007-3024-3_1

Leigh, J. P., & Schembri, M. (2004). Instrumental variables technique: Cigarette price provided better estimate of effects of smoking on SF-12. *Journal of Clinical Epidemiology, 57*(3), 284–293. https://doi.org/10.1016/j.jclinepi.2003.08.006

Leuven, E., & Sianesi, B. (2018). *PSMATCH2: Stata module to perform full Mahalanobis and propensity score matching, common support graphing, and covariate imbalance testing.* EconPapers. https://EconPapers.repec.org/RePEc:boc:bocode:s432001

Lu, M. (1999). The productivity of mental health care: An instrumental variable approach. *Journal of Mental Health Policy and Economics, 2*(2), 59–71. https://doi.org/10.1002/(SICI)1099-176X(199906)2:2<59::AID-MHP47>3.0.CO;2-J

Lumley, T. (2020). *survey: Analysis of complex survey samples.* https://CRAN.R-project.org/package=survey

Lunt, M. (2013). Selecting an appropriate caliper can be essential for achieving good balance with propensity score matching. *American Journal of Epidemiology, 179*(2), 226–235. https://doi.org/10.1093/aje/kwt212

MacKinnon, D. (2012). *Introduction to statistical mediation analysis.* Lawrence Erlbaum. https://doi.org/10.4324/9780203809556

Maddala, G. S., & Lahiri, K. (1992). *Introduction to econometrics* (Vol. 2). Macmillan.

Martens, E. P., Pestman, W. R., de Boer, A., Belitser, S. V., & Klungel, O. H. (2006). Instrumental variables: Application and limitations. *Epidemiology, 17*(3), 260–267. https://doi.org/10.1097/01.ede.0000215160.88317.cb

Maxwell, S. E. (2010). Introduction to the special section on Campbell's and Rubin's conceptualizations of causality. *Psychological Methods, 15*(1), 1–2. https://doi.org/10.1037/a0018825

Maxwell, S. E., & Delaney, H. D. (1993). Bivariate median splits and spurious statistical significance. *Psychological Bulletin, 113*(1), 181–190. https://doi.org/10.1037/0033-2909.113.1.181

McBee, M., & Field, S. (2017). Confirmatory study design, data analysis, and results that matter. In M. C. Makel & J. Plucker (Eds.), *Toward a more perfect psychology: Trust, accuracy, transparency* (pp. 59–78). American Psychological Association. https://doi.org/10.1037/0000033-004

Meade, A. W., Lautenschlager, G. J., & Hecht, J. E. (2005). Establishing measurement equivalence and invariance in longitudinal data with item response theory. *International Journal of Testing, 5*(3), 279–300. https://doi.org/10.1207/s15327574ijt0503_6

Menard, S. (2002). *Applied logistic regression analysis.* Sage. https://doi.org/10.4135/9781412983433

Mersmann, O., Trautmann, H., Steuer, D., & Bornkamp, B. (2018). *truncnorm: Truncated normal distribution.* https://CRAN.R-project.org/package=truncnorm

Miguel, E., Satyanath, S., & Sergenti, E. (2004). Economic shocks and civil conflict: An instrumental variables approach. *Journal of Political Economy, 112*(4), 725–753. https://doi.org/10.1086/421174

Ming, K., & Rosenbaum, P. R. (2001). A note on optimal matching with variable controls using the assignment algorithm. *Journal of Computational and Graphical Statistics, 10*(3), 455–463. https://doi.org/10.1198/106186001317114938

Moisen, G. (2008). Classification and regression trees. In S. E. Jørgensen & B. D. Fath (Eds.), *Encyclopedia of ecology* (Vol. *1*, pp. 582–588). Elsevier. https://doi.org/10.1016/B978-008045405-4.00149-X

Morgan, S. L., & Winship, C. (2014). *Counterfactuals and causal inference.* Cambridge University Press. https://doi.org/10.1017/CBO9781107587991

Mullahy, J. (1997). Instrumental-variable estimation of count data models: Applications to models of cigarette smoking behavior. *Review of Economics and Statistics, 79*(4), 586–593. https://doi.org/10.1162/003465397557169

Nosek, B. A., Ebersole, C. R., DeHaven, A. C., & Mellor, D. T. (2018). The preregistration revolution. *Proceedings of the National Academy of Sciences of the United States of America, 115*(11), 2600–2606. https://doi.org/10.1073/pnas.1708274114

Pearl, J. (1993). *Mediating instrumental variables* (Technical Report No. R-210). https://ftp.cs.ucla.edu/pub/stat_ser/r210.pdf

Pearl, J. (1995). Causal diagrams for empirical research. *Biometrika, 82*(4), 669–688. https://doi.org/10.1093/biomet/82.4.669

Pearl, J. (2009). *Causality: Models, reasoning, and inference.* Cambridge University Press.

Pearl, J. (2014). Interpretation and identification of causal mediation. *Psychological Methods, 19*(4), 459–481. https://doi.org/10.1037/a0036434

Pearl, J., Glymour, M., & Jewell, N. P. (2016). *Causal inference in statistics: A primer.* Wiley.

Pearl, J., & Mackenzie, D. (2018). *The book of why: The new science of cause and effect* (1st ed.). Basic Books.

Pei, Z., Lee, D. S., Card, D., & Weber, A. (2018). *Local polynomial order in regression discontinuity designs* (IRS Working Paper No. 622). Princeton University. https://dataspace.princeton.edu/bitstream/88435/dsp01v118rh27h/3/622.pdf

Pourhoseingholi, M. A., Baghestani, A. R., & Vahedi, M. (2012). How to control confounding effects by statistical analysis. *Gastroenterology and Hepatology from Bed to Bench, 5*(2), 79–83.

Rassen, J. A., Brookhart, M. A., Glynn, R. J., Mittleman, M. A., & Schneeweiss, S. (2009). Instrumental variables I: Instrumental variables exploit natural variation in nonexperimental data to estimate causal relationships. *Journal of Clinical Epidemiology, 62*(12), 1226–1232. https://doi.org/10.1016/j.jclinepi.2008.12.005

Rassen, J. A., Shelat, A. A., Myers, J., Glynn, R. J., Rothman, K. J., & Schneeweiss, S. (2012). One-to-many propensity score matching in cohort studies. *Pharmacoepidemiology and Drug Safety, 21*(Suppl. 2), 69–80. https://doi.org/10.1002/pds.3263

Raudenbush, S. W. (1997). Statistical analysis and optimal design for cluster randomized trials. *Psychological Methods, 2*(2), 173–185. https://doi.org/10.1037/1082-989X.2.2.173

Revelle, W. (2020). *psych: Procedures for psychological, psychometric, and personality research.* https://CRAN.R-project.org/package=psych

Ridgeway, G., McCaffrey, D., Morral, A., Griffin, B. A., Burgette, L., & Cefalu, M. (2020). *twang: Toolkit for weighting and analysis of nonequivalent groups.* https://CRAN.R-project.org/package=twang

Ripley, B., Venables, B., Bates, D. M., Hornik, K., Gebhardt, A., & Firth, D. (2020). *MASS: Support functions and datasets for Venables and Ripley's MASS.* https://CRAN.R-project.org/package=MASS

Robinson, D., Hayes, A., Couch, S., Patil, I., Chiu, D., Gomez, M., Demeshev, B., Menne, D., Nutter, B., Johnston, L., Bolker, B., Briatte, F., Arnold, J., Gabry, J., Selzer, L., Simpson, G., Preussner, J., Hesselberth, J., Wickham, H., . . . Wiernik, B. M. (2020). *broom: Convert statistical analysis objects into tidy tibbles.* https://CRAN.R-project.org/package=broom

Rohrer, J. M. (2018). Thinking clearly about correlations and causation: Graphical causal models for observational data. *Advances in Methods and Practices in Psychological Science, 1*(1), 27–42. https://doi.org/10.1177/2515245917745629

Rosenbaum, P. R., & Rubin, D. B. (1983). The central role of the propensity score in observational studies for causal effects. *Biometrika, 70*(1), 41–55. https://doi.org/10.1093/biomet/70.1.41

Rubin, D. B. (2005). Causal inference using potential outcomes: Design, modeling, decisions. *Journal of the American Statistical Association, 100*(469), 322–331. https://doi.org/10.1198/016214504000001880

Saltelli, A., Tarantola, S., Campolongo, F., & Ratto, M. (2004). *Sensitivity analysis in practice: A guide to assessing scientific models.* Wiley.

Schapire, R. E., & Freund, Y. (2012). *Boosting: Foundations and algorithms* (T. Dietterich, Ed.). MIT Press.

Scott, D. W. (2015). *Multivariate density estimation: Theory, practice, and visualization* (2nd ed.). Wiley. https://doi.org/10.1002/9781118575574

Sekhon, J. S. (2020). *Matching: Multivariate and propensity score matching with balance optimization.* https://CRAN.R-project.org/package=Matching

Shadish, W. R. (2010). Campbell and Rubin: A primer and comparison of their approaches to causal inference in field settings. *Psychological Methods, 15*(1), 3–17. https://doi.org/10.1037/a0015916

Shadish, W. R., Cook, T. D., & Campbell, D. T. (2001). *Experimental and quasi-experimental designs for generalized causal inference.* Houghton Mifflin.

Shadish, W. R., Cook, T. D., & Campbell, D. T. (2002). *Experimental and quasi-experimental designs for generalized causal inference.* Wadsworth Cengage Learning.

Shaughnessy, J. J., Zechmeister, E. B., & Zechmeister, J. S. (2015). *Research methods in psychology* (10th ed.). McGraw-Hill.

Shrier, I., & Platt, R. W. (2008). Reducing bias through directed acyclic graphs. *BMC Medical Research Methodology, 8*, Article 70. https://doi.org/10.1186/1471-2288-8-70

Shrout, P. E., & Rodgers, J. L. (2018). Psychology, science, and knowledge construction: Broadening perspectives from the replication crisis. *Annual Review of Psychology, 69*, 487–510. https://doi.org/10.1146/annurev-psych-122216-011845

Silverman, B. W. (1986). *Density estimation for statistics and data analysis.* Chapman & Hall.

Simmering, J. (2014, March 19). Stop using bivariate correlations for variable selection. *R-bloggers.* https://www.r-bloggers.com/stop-using-bivariate-correlations-for-variable-selection/

Simonsohn, U., Simmons, J. P., & Nelson, L. D. (2019). *Specification curve: Descriptive and inferential statistics on all reasonable specifications.* SSRN. https://doi.org/10.2139/ssrn.2694998

Sobel, M. E. (2009). Causal inference in randomized and non-randomized studies: The definition, identification, and estimation of causal parameters. In R. E. Millsap & A. Maydeu-Olivares (Eds.), *The SAGE handbook of quantitative methods in psychology* (pp. 3–22). Sage. https://doi.org/10.4135/9780857020994.n1

Staiger, D., & Stock, J. (1997). Instrumental variables regression with weak instruments. *Econometrica, 65*(3), 557–586. https://doi.org/10.2307/2171753

Steegen, S., Tuerlinckx, F., Gelman, A., & Vanpaemel, W. (2016). Increasing transparency through a multiverse analysis. *Perspectives on Psychological Science, 11*(5), 702–712. https://doi.org/10.1177/1745691616658637

Stuart, E. A. (2010). Matching methods for causal inference: A review and a look forward. *Statistical Science, 25*(1), 1–21. https://doi.org/10.1214/09-STS313

Stukel, T. A., Fisher, E. S., Wennberg, D. E., Alter, D. A., Gottlieb, D. J., & Vermeulen, M. J. (2007). Analysis of observational studies in the presence of treatment selection bias: Effects of invasive cardiac management on AMI survival using propensity score and instrumental variable methods. *JAMA Journal of the American Medical Association, 297*(3), 278–285. https://doi.org/10.1001/jama.297.3.278

Textor, J., van der Zander, B., & Ankan, A. (2020). *dagitty: Graphical analysis of structural causal models.* https://CRAN.R-project.org/package=dagitty

Therneau, T., Atkinson, B., & Ripley, B. (2019). *rpart: Recursive partitioning and regression trees.* https://CRAN.R-project.org/package=rpart

Thissen, D., Steinberg, L., & Kuang, D. (2002). Quick and easy implementation of the Benjamini-Hochberg procedure for controlling the false positive rate in multiple comparisons. *Journal of Educational and Behavioral Statistics, 27*(1), 77–83. https://doi.org/10.3102/10769986027001077

Thistlethwaite, D. L., & Campbell, D. T. (1960). Regression-discontinuity analysis: An alternative to the ex post facto experiment. *Journal of Educational Psychology, 51*(6), 309–317. https://doi.org/10.1037/h0044319

Tipton, E. (2013). Improving generalizations from experiments using propensity score subclassification: Assumptions, properties, and contexts. *Journal of Educational and Behavioral Statistics, 38*(3), 239–266. https://doi.org/10.3102/1076998612441947

Trochim, W. M., Cappelleri, J. C., & Reichardt, C. S. (1991). Random measurement error does not bias the treatment effect estimate in the regression-discontinuity design: II. When an interaction effect is present. *Evaluation Review, 15*(5), 571–604. https://doi.org/10.1177/0193841X9101500504

Ulrich, R., & Wirtz, M. (2004). On the correlation of a naturally and an artificially dichotomized variable. *British Journal of Mathematical and Statistical Psychology, 57*(2), 235–251. https://doi.org/10.1348/0007110042307203

Urbanek, S. (2013). *png: Read and write PNG images.* https://CRAN.R-project.org/package=png

VanderWeele, T. J. (2009). Mediation and mechanism. *European Journal of Epidemiology, 24*(5), 217–224. https://doi.org/10.1007/s10654-009-9331-1

VanderWeele, T. J. (2019). Principles of confounder selection. *European Journal of Epidemiology, 34*(3), 211–219. https://doi.org/10.1007/s10654-019-00494-6

VanderWeele, T. J., & Shpitser, I. (2011). A new criterion for confounder selection. *Biometrics, 67*(4), 1406–1413. https://doi.org/10.1111/j.1541-0420.2011.01619.x

Venkataramani, A. S., Bor, J., & Jena, A. B. (2016). Regression discontinuity designs in healthcare research. *British Medical Journal, 352*(8050), Article i1216. https://doi.org/10.1136/bmj.i1216

West, S. G., Duan, N., Pequegnat, W., Gaist, P., Des Jarlais, D. C., Holtgrave, D., Szapocznik, J., Fishbein, M., Rapkin, B., Clatts, M., & Mullen, P. D. (2008).

Alternatives to the randomized controlled trial. *American Journal of Public Health,* *98*(8), 1359–1366. https://doi.org/10.2105/AJPH.2007.124446

Westfall, J., & Yarkoni, T. (2016). Statistically controlling for covariates is harder than you think. *PLOS ONE, 11*(3), Article e0152719. https://doi.org/10.1371/journal.pone.0152719

Westreich, D., & Greenland, S. (2013). The Table 2 fallacy: Presenting and interpreting confounder and modifier coefficients. *American Journal of Epidemiology, 177*(4), 292–298. https://doi.org/10.1093/aje/kws412

Wicherts, J. M., Veldkamp, C. L., Augusteijn, H. E., Bakker, M., van Aert, R. C. M., & van Assen, M. A. L. M. (2016). Degrees of freedom in planning, running, analyzing, and reporting psychological studies: A checklist to avoid *p*-hacking. *Frontiers in Psychology, 7,* Article 1832. https://doi.org/10.3389/fpsyg.2016.01832

Wickham, H. (2020). *reshape2: Flexibly reshape data: A reboot of the reshape package.* https://CRAN.R-project.org/package=reshape2

Wickham, H., Chang, W., Henry, L., Pedersen, T. L., Takahashi, K., Wilke, C., Woo, K., Yutani, H., Dunnington, D., & RStudio. (2020). *ggplot2: Create elegant data visualisations using the grammar of graphics.* https://CRAN.R-project.org/package=ggplot2

Wickham, H., François, R., Henry, L., Müller, K., & RStudio. (2020). *dplyr: A grammar of data manipulation.* https://CRAN.R-project.org/package=dplyr

Wickham, H., & RStudio. (2020). *tidyr: Tidy messy data.* https://CRAN.R-project.org/package=tidyr

Wilke, C. O. (2020). *cowplot: Streamlined plot theme and plot annotations for "ggplot2".* https://CRAN.R-project.org/package=cowplot

Wooldridge, J. M. (2016). *Introductory econometrics: A modern approach.* Nelson Education.

Xie, Y., Allaire, J. J., Kim, A., Samuel-Rosa, A., Oles, A., Yasumoto, A., Frederik, A., Quast, B., Marwick, B., Ismay, C., Dervieux, C., Franklund, C., Emaasit, D., Shuman, D., Attali, D., Tyre, D., Valentiner, E., van Dunne, F., Wickham, F., . . . FriendCode. (2020). *bookdown: Authoring books and technical documents with R markdown.* https://CRAN.R-project.org/package=bookdown

Xie, Y., Vogt, A., Andrew, A., Zvoleff, A., Simon, A., Atkins, A., Wolen, A., Manton, A., Yasumoto, A., Baumer, B., Diggs, B., Zhang, B., Pereira, C., Dervieux, C., Hugh-Jones, D., Robinson, D., Hemken, D., Murdoch, D., Campitelli, E., . . . Foster, Z. (2020). *knitr: A general-purpose package for dynamic report generation in R.* https://CRAN.R-project.org/package=knitr

Yarkoni, T., & Westfall, J. (2017). Choosing prediction over explanation: Lessons from machine learning. *Perspectives on Psychological Science, 12*(6), 1100–1122. https://doi.org/10.1177/1745691617693393

INDEX

Page numbers in *italic* indicate figures and in **bold** indicate tables.